Springer Series in

MATERIALS SCIENCE 74

Springer-Verlag Berlin Heidelberg GmbH

Springer Series in

MATERIALS SCIENCE

The Springer Series in Materials Science covers the complete spectrum of materials physics, including fundamental principles, physical properties, materials theory and design. Recognizing the increasing importance of materials science in future device technologies, the book titles in this series reflect the state-of-the-art in understanding and controlling the structure and properties of all important classes of materials.

61 **Fatigue in Ferroelectric Ceramics and Related Issues**
By D.C. Lupascu

62 **Epitaxy**
Physical Principles and Technical Implementation
By M.A. Herman, W. Richter, and H. Sitter

63 **Fundamentals of Ion-Irradiated Polymers**
By D. Fink

64 **Morphology Control of Materials and Nanoparticles**
Advanced Materials Processing and Characterization
Editors: Y. Waseda and A. Muramatsu

65 **Transport Processes in Ion-Irradiated Polymers**
By D. Fink

66 **Multiphased Ceramic Materials**
Processing and Potential
Editors: W.-H. Tuan and J.-K. Guo

67 **Nondestructive Materials Characterization**
With Applications to Aerospace Materials
Editors: N.G.H. Meyendorf, P.B. Nagy, and S.I. Rokhlin

68 **Diffraction Analysis of the Microstructure of Materials**
Editors: E.J. Mittemeijer and P. Scardi

69 **Chemical-Mechanical Planarization of Semiconductor Materials**
Editor: M.R. Oliver

70 **Applications of the Isotopic Effect in Solids**
By G.V. Plekhanov

71 **Dissipative Phenomena in Condensed Matter**
Some Applications
By S. Dattagupta and S. Puri

72 **Predictive Simulation of Semiconductor Processing**
Status and Challenges
Editors: J. Dabrowski and E.R. Weber

73 **SiC Power Materials and Devices**
Some Applications
Editor: Z.C. Feng

74 **Plastic Deformation in Nanocrystalline Materials**
By M.Yu. Gutkin and I.A. Ovid'ko

75 **Wafer Bonding**
Applications and Technology
Editors: M. Alexe and U. Gösele

76 **Spirally Anisotropic Composites**
By G.E. Freger, V.N. Kestelman, and D.G. Freger

Volumes 10–60 are listed at the end of the book.

M.Yu. Gutkin   I.A. Ovid'ko

Plastic Deformation in Nanocrystalline Materials

With 69 Figures

Springer

Dr. Mikhail Yu. Gutkin
Prof. Ilya A. Ovid'ko
Institute of Problems of Mechanical Engineering
Russian Academy of Sciences
Bolshoy 61, Vasil. Ostrov
St. Petersburg 199178, Russia
E-mail: gutkin@def.ipme.ru
ovidko@def.ipme.ru

Series Editors:

Professor Robert Hull
University of Virginia
Dept. of Materials Science and Engineering
Thornton Hall
Charlottesville, VA 22903-2442, USA

Professor R. M. Osgood, Jr.
Microelectronics Science Laboratory
Department of Electrical Engineering
Columbia University
Seeley W. Mudd Building
New York, NY 10027, USA

Professor Jürgen Parisi
Universität Oldenburg, Fachbereich Physik
Abt. Energie- und Halbleiterforschung
Carl-von-Ossietzky-Strasse 9–11
26129 Oldenburg, Germany

Professor Hans Warlimont
Institut für Festkörper-
und Werkstofforschung,
Helmholtzstrasse 20
01069 Dresden, Germany

ISSN 0933-033X
ISBN 978-3-642-05903-2 ISBN 978-3-662-09374-0 (eBook)
DOI 10.1007/978-3-662-09374-0

Library of Congress Cataloging-in-Publication Data applied for.
Bibliographic information published by Die Deutsche Bibliothek
Die Deutsche Bibliothek lists this publication in the Deutsche Nationalbibliografie;
detailed bibliographic data is available in the Internet at http://dnb.ddb.de.

springeronline.com

Originally published by Springer-Verlag Berlin Heidelberg New York in 2004
Softcover reprint of the hardcover 1st edition 2004

Typesetting by the authors
Cover concept: eStudio Calamar Steinen
Cover production: *design & production* GmbH, Heidelberg

Printed on acid-free paper SPIN: 10978245 57/3141/tr 5 4 3 2 1 0

Preface

It seems there is no special need to comment on the term 'nanostructure' now, when one often meets the 'nano' words not only in scientific journals but even in newspapers. Moreover, today they are even to be heard in TV and radio programmes. In academic science, where the terms 'nanostructure' and 'nanotechnology' have been extremely popular since the early 1990s, they have been successfully extended to the sphere of economics and business, and now to politics. This is quite natural because nanostructures and nanotechnologies will surely serve as a basis for the most advanced and highest technology production in the nearest and probably also the remote future. Hence, the struggle to create and occupy its markets is already under way.

In this respect, it is of great interest to review data on the dynamics of U.S. Federal Goverment expenditure for nanotechnology [1, 2]. In the fiscal years 1997 and 2002, expenditure was approximately US$ 116 and US$ 697 million, respectively. In the fiscal year 2004, the President's request for US federal investment in nanoscale science, engineering and technology is about US$ 849 million [2]. The indicative budget allocated to the Thematic Priority entitled 'Nanotechnologies and nanosciences, knowledge-based multifunctional materials and new production processes and devices' for the duration 2002–2006 of the sixth EU Framework Programme for Research and Technological Development is EUR 1300 million [3]. The importance and priority of nanotechnologies is supported by worldwide efforts in fundamental and applied research as well as intense development of nanobusiness. In 2001, the annual turnover of the world nanoproduction market was approximately US$ 45 billion and after 10–15 years, it is expected to reach about US$ 1 trillion [4]. Even the more reserved prediction of the British Institute of Nanotechnology that the annual turnover of the global market for nanoproducts in 2010 may achieve US$ 145 billion [5] demonstrates the extremely dynamic development in this sphere.

Among various solid-state nanostructures, nanocrystalline materials (NCMs) attract special attention due to their outstanding mechanical properties, many of which have not been well understood or explained until now. Our desire to consider different ideas and present our own view on this topic has become the main motivation for writing this book.

The book consists of six chapters. Chapter 1 serves as an introduction, in which we consider different types of deformed solid-state nanostructure and discuss some common features in their structure and mechanical behavior, paying special attention to the role of such defects as dislocations, disclinations and grain boundaries.

Chapter 2 is devoted to theoretical models of plastic deformation in nanocrystalline metals, alloys and ceramics. The main aim is to trace the relations between the structural peculiarities and mechanisms of plasticity. We review the work which has been done to explain the anomalous dependence of the yield stress on the grain size (Hall–Petch effect) in NCMs. Special attention is paid to models which account for the grain size distribution, triple junction diffusion and grain boundary sliding.

In Chap. 3, we deal with the models of plastic flow localization in NCMs. Starting from the concepts of cooperative grain boundary sliding and cellular dislocations, we discuss the description of the latter with the help of elastic solutions obtained within the gradient theory of elasticity and analyse models of localized (super)plasticity in nanocrystalline ceramic films under the action of indenter. Theoretical schemes for cellular dislocation generation and propagation under conditions of low and high intensity of accommodation processes are proposed. The possible relation of low (high) intensity accommodation processes with the homogeneous (heterogeneous) regime of plastic deformation is demonstrated. For a model ensemble of cellular dislocations, we suggest a system of coupled chemical-reaction-kinetics differential equations and give an exact solution in an explicit form. The solution may further be used to model cellular dislocation behavior under a wide range of experimental conditions. Considering deformation processes on the nanoscopic level, we also discuss a model that describes both the strengthening and softening of NCMs under superplastic deformation due to the effects of triple junctions of grain boundaries as obstacles for grain boundary sliding. Within the framework of the model, the dependence of the yield stress for grain boundary sliding on defect and triple junction parameters is revealed and compared with experimental data.

Chapter 4 discusses different models of rotational plastic deformation in nano- and polycrystalline materials. In the first part of the chapter, we consider disclination models of generation and development of misorientation bands in severely deformed metals and alloys. The models predict the existence of a critical external shear stress, above which nucleation of misorientation bands takes place. Further analysis demonstrates two main regimes of misorientation band development, viz., stable and unstable propagation, and allows one to find another critical stress that controls the transition between these two regimes. We also quote some results from computer simulations of the 2D dynamics of dislocations in the stress field of a dipole of partial wedge disclinations in order to elucidate the micromechanisms of misorientation band propagation. The second part of the chapter is devoted to the theoretical models of grain boundary disclination motion which leads to changes in misorientation across grain boundaries in NCMs and may explain the rotation of the nanograin crystalline lattice as a whole. It is demonstrated that motion of grain boundary disclinations may occur in NCMs through emission of pairs of lattice dislocations into the adjacent grains or through climb of grain boundary dislocations. We consider a model of crossover from grain boundary sliding to rotational deformation which is realized by the transformation of a pile-up of gliding grain boundary dislocations stopped by a triple junction of grain boundaries, into two walls of climbing grain boundary dislocations (treated as the dipoles of partial wedge disclinations). The conditions necessary for such a transformation are determined and discussed.

Chapter 5 considers some possible ways for splitting the triple junction and grain boundary disclinations into disclinations of smaller strength, as well as the related problems of solid-state amorphization and micro- and nanocrack generation. In particular, a model of solid-state amorphization of grain boundary triple junctions which accompanies the process of splitting of a triple junction disclination is described. The critical strength of the disclination at which the splitting becomes energetically favourable is determined. It is shown that such a process competes effectively with microcrack generation at triple junctions of grain boundaries, causing a plastification of deformed polycrystalline materials. We also analyse nanocracks that are nucleated at triple junction or grain boundary disclinations and propagate along grain boundaries, changing the direction of their growth at triple junctions of grain boundaries.

In the last short chapter, we give the main conclusions and consider the prospects for further developments in the theory of plastic deformation in nanocrystalline materials.

We would like to express our warmest thanks to a great many friends and colleagues for helpful discussions and collaboration. These include E.C. Aifantis, S.V. Bobylev, A.A. Fedorov, P. Klimanek, A.L. Kolesnikova, R.A. Masumura, K.N. Mikaelyan, N.F. Morozov, C.S. Pande, A.B. Reizis, A.E. Romanov, M. Seefeldt, A.G. Sheinerman and N.V. Skiba.

This work was supported, in part, by the Office of US Naval Research (Grants N00014-99-1-0569, N00014-99-1-0896 and N00014-01-1-1020), the Russian Fund of Basic Research (Grants 98-02-16075 and 01-02-16853), the Russian Academy of Sciences Program entitled 'Structural Mechanics of Materials and Constructions', St. Petersburg Scientific Center of the Russian Academy of Sciences, INTAS (Grant 99-01216), the Volkswagen Foundation (Research Project 05019225), the Russian State Research Program on Solid-State Nanostructures, the Russian Science Support Foundation, and 'Integration' Program (Grant B0026).

St. Petersburg,
January 2004

Mikhail Gutkin
Ilya Ovid'ko

Contents

1 Introduction

1.1 Deformed Solid-State Nanostructures

The present book deals with plastic deformation processes in the so-called nanocrystalline materials, a special type of deformed solid-state nanostructure which comprises a significant fraction of the general class of nanostructures. Typical examples of deformed solid-state nanostructures are bulk nanocrystalline metals and alloys, nanolayered heterostructures, nanocrystalline thin films and coatings, etc. [6–8]. The most common feature of these structures is that they consist of small particles or layers which are separated by specific interfaces and have a characteristic size from 1 to 100 nanometers (1 nm $= 10^{-9}$ m). The bottom limit of this range corresponds to some 2–4 interatomic distances in a solid, while the top limit is quite symbolic. In our previous book [8], we subdivided solid-state nanostructures into the ordered and disordered ones. For example, bulk nanocrystalline metals and alloys as well as nanocrystalline thin films may be classified as disordered nanostructures because the size of their nanograins (nanocrystallites) is a random quantity, not to mention the randomness of their shape, inhomogeneity in the chemical composition and defect structure across the sample. In contrast, due to peculiarities in their fabrication and function, high-quality nanolayered heterostructures may be classified as ordered nanostructures.

Most advanced disordered nanostructures appeared relatively recently and became the subject of intensive investigation at the end of the 1980s. The first nanocrystalline materials were obtained in the early 1980s. They are polycrystalline solids with grain size ranging from 3 to 100 nm [1,6–35]. At such small grain sizes, the volume fraction of grain boundaries achieves very high values, in which case such boundaries crucially affect macroscopic properties of a nanocrystalline solid. An intermediate position between nanocrystalline metals and metallic glasses is occupied by amorphous-nanocrystalline composites which were distinguished as a special class of materials in 1987 [36–40]. They consist of two phases: the amorphous phase plays the role of the matrix while the nanocrystalline phase is represented by separate nanocrystallites which look like inclusions. It is important that the mechanical properties (microhardness, yield stress, strength) of such composites may significantly (several times) exceed those of the corresponding material in a purely amorphous or a conventional polycrystalline state. At the same time, the ductility of these composites remains rather high, much higher than in the amorphous state [36].

Both the nanocrystalline solids and amorphous-nanocrystalline composites are metastable, inhomogeneous and disordered systems with the characteristic nanoscopic scale of inhomogeneity. The physical mechanisms responsible for their mechanical properties have been poorly studied to date. This is mainly explained by the fact that we are still a long way from understanding the defect behavior in such nanostructures. One can only affirm that this behavior significantly differs from the behavior of defects in conventional materials. It is worth noting the important role of interfaces and their triple junctions and disclinations, all of which may be treated as integral elements of the structure in nanomaterials (i.e., so-called 'structural' defects). We have already considered the main peculiarities of defect behavior in nanovolumes in our previous book [8]. They will also be listed briefly in the next section because it seems to be impossible to model and discuss the mechanical behavior of nanostructures without accounting for these peculiarities.

As an example of the ordered inhomogeneous nanostructures, one can take the so-called 'strained layered superlattices' (SLS), one of the most advanced types of solid thin-film system which has been developed dynamically since the end of the 1980s [41–48]. SLSs are multilayered (sometimes isoperiodic) structures of alternate films having different chemical compositions and nanoscopic thicknesses. They were first proposed as novel semiconductor heterostructures with unique electronic and optical properties [41,42]. Later they were investigated to study the mechanisms of misfit dislocation generation and determine the corresponding critical values for the SLS parameters [43–45]. Nowadays SLSs are widely used in various fields of micro- and optoelectronics [46], in particular, as special buffer nanostructures introduced into semiconductor thin-film systems to clean them from threading dislocations (e.g., see [47,48] and references therein). In the last decade, metallic and ceramic SLSs have attracted much attention as novel nanostructures with unusual mechanical, electrical and magnetic properties [1,2,20–27,29–32,49–53].

Theoretical description of the mechanical behavior of contemporary nanocrystalline and nanostructured materials is still in its infancy. Some basic approaches have been formulated, and models of the most interesting and astonishing phenomena are being elaborated.

In particular, much attention has been paid to the development of elastic models for defects in nanoparticles, nanocrystalline solids and thin-film nanostructures, which require the solution of new and often very complicated and cumbersome boundary-value problems in the classical theory of elasticity (see the reviews in [8,15,54–62] and original papers [63–140]).

Much effort is also being made to describe defects within non-classical theories such as the nonlocal theory of elasticity [141–148], the gradient theory of elasticity [149–158] and the gauge theory of elastoplasticity [159–167]. All these allow one to dispense with the classical singularities of elastic fields of defects (dislocations and disclinations) at their lines and also to study the short-range elastic interactions of defects with each other [149, 150, 152, 153, 158, 164] and with interphase boundaries [144, 155–158]. It is worth noting that many solutions obtained using these quite different theories coincide totally. For example, the elastic stress fields of screw dislocations are given by the same formulas in the nonlocal [141–143], gradient [151, 154, 158] and gauge field [159–161, 163, 164] theories. The gradient solutions for the stresses, strains and energies of both the screw and edge dislocations [149–151,154,158] coincide with those in the gauge theory [163–165]. The elastic fields of both the wedge and twist disclinations also turn out the same in the nonlocal [147], gradient [152–154, 158] and gauge [166, 167] theories. Such nonsingular solutions allow one to describe the behavior of defects in the vicinity of the outer and interphase boundaries [155–158] where the elastic moduli suffer jumps, and in nanosolids whose sizes are of the same order of magnitude as the size of the defect core [164–166]. Application of these new results to the theory of the mechanical behavior of nanostructures is still an open question. There is only one model and this will be considered in Chap. 3 of the present book.

In general, the mechanics of deformed nanostructures is now being developed in several directions, namely, those that seem most necessary today. At the same time, the total amount of relevant information is already enormous and growing all the time. Hence, in writing this book, we concentrate mainly on results that have been obtained with our own participation, as well as similar results from other authors.

1.2 Peculiarities in the Structure of Nanocrystalline Materials

Extensive investigations of structure, properties and fabrication methods of nanocrystalline materials (NCMs) which have been carried out over the last two decades [1, 2, 8–40, 49–53, 168–201] have shown that NCMs differ qualitatively from conventional polycrystals. In the first place, this concerns the structure of grain (intercrystalline) boundaries whose thickness in NCMs may reach 1–2 nm. This means that the grain boundaries themselves are typical nano-objects – layers of material which often have another atomic structure and are sometimes much more porous than the material inside the grains (crystallites). Comparing experimental data from Mössbauer spectroscopy [168], positron lifetime spectroscopy [169, 170], X-ray diffraction [171], EXAFS [175], neutron diffraction [177] and HREM [181], one can conclude that a significant part of the grain boundaries in NCMs have severely distorted near-boundary regions with lower atomic density and a higher level of elastic strain [181]. Microstructural studies [181] of nanocrystalline palladium with grain size 4–9 nm, which was fabricated by high-pressure compaction of nanocrystallites condensed from the gas phase, demonstrate that it contains $\approx$ 40% vol. undistorted crystalline material, $\approx$ 25% stretched or amorphous-type grain boundary layers, $\approx$ 25% highly strained material, and $\approx$ 10% pores. Inside the grains, lamellae of twins, low-angle boundaries and dislocations localized near the grain boundaries have been observed. Under thermal annealing, this system was transformed into a conventional polycrystalline structure with thin grain boundaries and undistorted grains.

Besides the grain boundaries, their triple junctions play an important role in the behavior of NCMs. Indeed, if the mean grain size is roughly several nanometers and the grain boundary thickness equals 1–2 nm, then the volume fraction of the triple junction material is very high in such NCMs. In recent years, it has been clearly recognized that triple junctions of grain boundaries have different structure and properties from the grain boundaries that they adjoin [202]. From experimental data and theoretical models [180, 202–211], it follows that the triple junctions act as enhanced diffusivity tubes, nuclei of second phase segregation, strengthening elements and sources of lattice dislocations during plastic deformation, and drag centers of grain boundary migration during recrystallization processes. In particular, the outstanding diffusional properties exhibited by NCMs [33, 212–215] are viewed as being related to the effect of the highly enhanced diffusion along triple junction tubes [10].

Experimental research into lattice defects in NCMs are often based on HREM observations. These give images that need special treatment because many image details bear no relation to the real structure of the NCM [195]. For this reason, most papers aiming to study atomic structure in direct resolution are now accompanied by computer simulations of the atomic structure, and the results are represented as superpositions of the HREM and simulated images. In the case of defect structures in NCMs, such studies are still quite rare. However, there have been a number of HREM observations of defects in various nanostructures, viz., small particles and whiskers [15, 55, 56, 195], NCMs [15, 181, 189, 195, 196, 198, 216–219] and strained layered superlattices [220–223].

On the basis of data gathered from the literature, the following main peculiarities of defect structures in NCMs may be highlighted [8]:

- All defects in NCMs can be subdivided conventionally into structural and induced defects. Structural defects cannot be separated from the structure of the material. Their elimination would lead to a change in the 'ideal' structure of the solid (the 'physical vacuum') when the whole system transforms into a qualitatively new state. These are, for example, the wedge disclinations in the disclination models of small particles [15, 55, 56, 59, 87, 88, 90, 91], grain boundaries and their triple junctions which also often contain the disclinations in NCMs [9–16, 33, 56, 60, 62, 168–211, 216–219], and interfaces between the amorphous and crystalline phases in the amorphous-nanocrystalline composites [36–40]. The induced defects appear as a response of the system to changes in external conditions (temperature, mechanical load, chemical action of the environment, etc.) and provide the relaxation processes, e.g., glide and climb of dislocations, diffusional mass transfer, formation of twin lamellae, shear and misorientation bands, etc.

- In such disordered nanostructures as the NCMs and amorphous-nanocrystalline composites, the structural defects form disordered arrangements, in contrast to their ordered arrangements in ordered nanostructures like strained layered superlattices.
- Construction and properties of structural defects are determined by the conditions of NCM fabrication (as follows directly from their definition) as well as by the size effect. For example, if the mean size d of nanocrystallites in an NCM is close to the lower limit of the interval 1–100 nm, then the intercrystallite boundaries may have a non-equilibrium structure [185,187, 191,197] which transforms into the equilibrium structure under further thermal annealing, accompanied by growth of the nanocrystallites. In the opposite case, when d is closer to the upper limit of this interval, most of the intercrystallite boundaries have normal equilibrium structure [198].
- As a rule, an individual unloaded nanovolume contains few induced defects, normally localized just near its boundaries, or else it does not contain them at all. This explains the enhanced interest in studying the conditions of stability of induced defects (dislocations and dislocation loops) in such systems. The smaller nanovolumes contain fewer induced defects. For any material, one can estimate the critical size d_c of a nanovolume (here a particle with free surface) such that, for $d < d_c$, the probability of induced defects occurring in it drastically decreases [15,56,59,77,78,86], and $d_c \approx$ 10–100 nm [77]. For the model of a nanocrystalline solid with slipping intercrystalline boundaries, theoretical estimates of d_c vary from 2–3 nm (for α-Fe) to 24–38 nm (for Cu) [86]. These estimates are in good agreement with experimental observations [181,189,196,198]. As an exception, we can mention the induced defects of growth type, whose generation is controlled by the conditions of nanostructure fabrication (growth). For example, the density of threading dislocations in the first (bottom) layers of strained superlattices is determined by their density in the substrate [44–47].
- The mechanisms of plastic deformation and fracture in NCMs are also determined in great part by the size effect. For example, Ke et al. [189,196] observed in situ how, at $d < 25$ nm, plastic deformation of nanocrystalline metallic films occurred through transformations of structural defects that were indicated by the phenomena of grain boundary sliding and rotation of nanocrystallites near the tip of an opening crack. At the same time, no induced defects were observed inside the nanocrystallites and the crack propagated along the grain boundaries through the nucleation of nanopores at the boundary triple junctions in front of the crack, diffusional growth of these nanopores and their merging with the crack tip.

At $d > 35$ nm, the crack was observed to propagate in brittle manner, with tearing of nanograins [189]. At $d \approx 110$ nm, the induced defects – individual dislocations – were fixed inside the nanograins [196]. Recently, Chen et al. [219] reported on deformation twinning in nanocrystalline pure aluminum, which is not typical for coarse-grained pure aluminum.

One can conclude from the peculiarities listed above that, in order to elaborate adequate theoretical models of deformational behavior in NCMs, one must start from models of structural defects and their evolution under external loading, analysis of the conditions for induced defect generation and stability, and knowledge of interaction mechanisms between structural and induced defects. In the following chapters, we will try to show how the structural peculiarities of NCMs are taken into account in theoretical considerations of different aspects of their plastic deformation.

2 Yield Stress of Nanocrystalline Materials

Most mechanical properties of nanocrystalline materials (NCMs) are essentially different from those of conventional coarse-grained polycrystals. NCMs exhibit extremely high strength and good fatigue resistance [224–226], as desired for numerous applications. Most NCMs show low tensile ductility which limits their practical utility. At the same time, several examples of substantial ductility and even superplasticity have been reported. In particular, nanocrystalline ceramics commonly exhibit superplasticity at lower temperatures and faster strain rates than their coarse-grained counterparts [227, 228].

Specific features of deformation processes in NCMs manifest themselves in deviations from the known grain size scaling relations. The classical Hall–Petch law [229, 230] describes the relationship between yield stress τ and grain size d of a polycrystalline material, viz.,

$$\tau = \tau_0 + Kd^{-1/2}\,, \tag{2.1}$$

where τ_0 is the friction stress needed to move individual dislocations, and K is a (material dependent) constant, often referred to as the Hall–Petch slope. This equation is well behaved for grains larger than about a micron. Masumura et al. [231] plotted some of the available data in a Hall–Petch plot $(\tau - \tau_0)/K$ vs. $d^{-1/2}$. This demonstrates that the yield stress–grain size exponent for relatively large grains appears to be very close to -0.5, and this trend generally continues until the very fine grain regime (~ 100 nm) is reached. With the advent of NCMs whose grain sizes are of nanometer (nm) dimensions, the applicability and validity of (2.1) becomes of interest in view of recently compiled experimental results [232].

A close analysis of experimental Hall–Petch data in a variety of materials shows three different regions:

- a region from a single crystal to a grain size of about a micron, where the classical Hall–Petch description can be used;
- a region for grain sizes ranging from about a micron to about 30 nm, where the Hall–Petch relation holds roughly, but deviates from the classical -0.5 exponent to a value near zero;
- a region beyond a very small critical grain size where the Hall–Petch slope is nearly zero, with no increase in strength on decreasing grain size or where the strength actually decreases with decreasing grain size.

Although some of the measurements on which the trend discussed above is based are not entirely reliable, for a variety of reasons discussed recently by Sanders et al. [233], the above delineation into three regions is beginning to be accepted.

The specific peculiarities of the grain size dependence of the yield stress in NCMs are definitely caused by their structural peculiarities (see Sect. 1.2), the main ones being the nanoscale structure and the extremely high volume fraction of the grain boundary phase. At the same time, NCMs are difficult to characterize experimentally because their structural units – nanograins – are much smaller than visible wavelengths and larger than individual molecules. In this context, an adequate theoretical description of the grain size dependence of the yield stress in NCMs is, on the one hand, capable of essentially contributing to understanding the fundamentals of structure–property relationships in nanoscale solids and will, on the other hand, serve as a basis for development of high technologies exploiting the outstanding mechanical properties of NCMs.

At present, the nature of deviations from the conventional Hall–Petch grain size scaling relation in NCMs is a subject of controversy. In general, theoretical models of this phenomenon can be divided into two basic categories:

- models describing NCMs as two-phase composites with grain interiors and boundaries playing the role of component phases;
- models describing the evolution of defects and grain boundary structures, with focus placed on the specific features of physical mechanisms (lattice dislocation motion, grain boundary sliding, diffusion plasticity mechanisms) of plastic flow in NCMs.

Models in the first category operate with the averaged mechanical characteristics and volume fractions of the grain-interior and grain-boundary phases. Models in the second category deal with nanoscale effects on conventional lattice dislocation motion, competition between various deformation mechanisms, and the effects of a distribution of grain size on this competition in NCMs. The main aim of this chapter is to provide an overview of the theoretical models of plastic deformation processes in NCMs paying particular attention to the abnormal Hall–Petch relationship.

Let us briefly characterize the basic aspects of theoretical models describing the grain size dependence of the yield stress in NCMs. All the models exploit the key structural peculiarities of NCMs – their nanoscale structure and the high volume fraction of the grain boundary phase – as input in analysis of the deformation behavior of such materials. In doing so, models in the different categories use different theoretical methods to describe plastic deformation and distinguish different structural peculiarities as playing the crucial role in plastic deformation processes in NCMs.

Several models exploit the idea of lattice dislocation motion in grain interiors as the basic deformation mechanism in NCMs, which is modified (com-

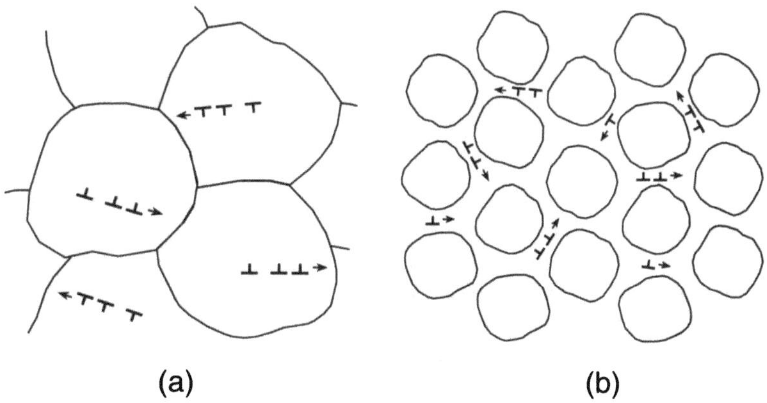

Fig. 2.1. Grain boundaries (**a**) play the role of obstacles for lattice dislocations in coarse-grained polycrystals and (**b**) serve as softening structural elements that carry the plastic flow in nanocrystalline materials

pared to that in coarse-grained polycrystals) due to nanoscale effects. At the same time, the generic idea of most of the models is that the grain boundary phase provides the effective action of the deformation mechanism(s) in NCMs, which is (are) different from the lattice dislocation mechanism realized in conventional coarse-grained polycrystals. In the context of this generic idea, the dominant role of the lattice dislocation mechanism causes the grain refinement to strengthen a coarse-grained material (Fig. 2.1a), in which case the classical Hall–Petch relationship is valid. At the same time, when the deformation mechanisms associated with the active role of grain boundaries effectively come into play (Fig. 2.1b), the grain refinement will weaken a specimen. This is the case for NCMs. The theoretical models under consideration are distinguished by their identification of the deformation mechanism(s) inherent in grain boundaries and their description of the competition between these mechanisms and the conventional lattice dislocation motion. The following plastic deformation mechanisms are commonly treated as those associated with the active role of grain boundaries and the nanoscale structure, effectively competing with the lattice dislocation mechanism in NCMs: grain boundary sliding and diffusional mass transfer occurring predominantly via grain boundary diffusion.

Models of NCMs as two-phase composites (models in the first category above) describe the yield stress τ and other mechanical characteristics with the help of the so-called rule of mixture. In this approach, the yield stress τ of an NCM is some weighted sum (mixture) of the yield stresses characterizing the grain-interior and grain-boundary phases, which strongly depends on the volume fraction of the grain boundary phase and, therefore, the grain size d. The yield stress of the grain-boundary phase is assumed to be lower than that of the grain-interior phase, in which case the rule of mixture describes the deviations from the conventional Hall–Petch relationship in accordance with experimental data. Such models do not usually take into account the evolution of defects and transformations of grain boundary structures, which, generally speaking, are capable of strongly influencing the deformation behavior of a mechanically loaded NCM. However, the models in question are based on the indirect assumption that grain boundaries cause the effective action of the plastic deformation mechanism(s), which is (are) different from the conventional lattice dislocation motion inherent in the bulk crystalline phase. The theoretical models based on the rule-of-mixture approach are reviewed in detail in Sect. 2.1.

The theoretical models in the second category directly address the physical mechanisms of plastic flow in NCMs and focus on the evolution of defects (lattice dislocations, grain boundary dislocations, vacancies) as carriers of plastic flow and transformations of grain boundary structures, which accompany deformation processes in NCMs. Such models often make a direct comparison between the characteristics of these mechanisms. That is, the plastic deformation mechanism that requires the least yield stress (or energy) is favoured, in which case the NCM is characterized by the yield stress inherent in the favoured deformation mechanism. A modified scheme for estimating the yield stress is suggested in the approach taking into account the distribution of grain sizes occurring in most nanocrystalline specimens. This approach exploits the idea of the simultaneous action of various deformation mechanisms in various local areas of a mechanically loaded NCM, depending on the grain size in such areas. In doing so, one operates with averaged (over a distribution of grain size) mechanical characteristics of NCMs. In this second category, we also include computer models describing physical mechanisms of plastic deformation in NCMs and transitions from one deformation mechanism to another, occurring with reduction in the grain size d. All the above models are considered in Sect. 2.2.

2.1 Rule-of-Mixture Approach to Yield Stress

The rule-of-mixture approach has become one of the most widespread ways to describe the Hall–Petch relation for nanostructured solids. The main idea of the simplest rule of mixture is, in order to estimate some macroscopic physical quantity which would describe a complex solid as a whole, to use an overall value equal to the sum of appropriate quantities for separate parts of the solid multiplied by volume fractions of the solid parts. More precisely, but more complicatedly, rules of mixture can take into account the geometry of the mutual distribution of components within the solid as well as some of their properties. Such theories of 'effective' overall macroscopic properties are mostly developed in the science of composite materials (e.g., see [234, 235]).

Kocks [236] was probably the first to use the analogy between conventional polycrystalline solids and composite materials. Indeed, disregarding mutual misorientations of grains, one can consider a polycrystalline solid as a composite which consists of a single-crystalline matrix and intercrystalline layers (inclusions). Assuming the bulk areas of grains (matrix) are characterized by the yield stress τ_{G} and the boundary layers of thickness t by the yield stress τ_{B}, Kocks [236] used the rule of mixture and obtained the estimate

$$\tau_{\mathrm{y}} = \tau_{\mathrm{G}} + \frac{4t}{d}\left(\tau_{\mathrm{B}} - \tau_{\mathrm{G}}\right) \tag{2.2}$$

for the yield stress τ_{y} of a polycrystalline solid with grain size d. Obviously, the Hall–Petch relation is not attributed to the model [236] because (2.2) gives $\tau_{\mathrm{y}} \sim d^{-1}$ instead of the standard Hall–Petch law, $\tau_{\mathrm{y}} \sim d^{-1/2}$. It is worth noting, however, that this is true until the quantities τ_{G} and τ_{B} are treated as constants which do not depend on the grain size d themselves. We will consider below some models where this assumption is not satisfied at all or else it is satisfied for τ_{B} alone. As noted in the review [237], Kocks' model [236] may only be correct for polycrystalline materials with very small grains. In fact, Jang and Koch [238], trying to find the source of the spread in the results they obtained, discovered that the microhardness of nanocrystalline iron may follow the d^{-1} dependence as well as the $d^{-1/2}$ dependence.

Kocks' idea of representing a polycrystalline material as a composite of a crystalline matrix with inclusions of intercrystalline layers was developed in the work by Gryaznov et al. [239], which aimed to obtain a generalized empirical analog of the Hall–Petch law and explain why it failed for small grain sizes. Some deviations of the $\tau_{\mathrm{y}}(d)$ dependence from the standard Hall–Petch law were already observed in early experiments on microhardness measurement for NCMs. For example, microhardness test experiments performed with nanocrystalline Ni [176,240], TiO_2 [173,174,241], Cu [176,242–244] and copper alloys [184], Pd [242–244], Co [176], Fe [194,238], Ag [190,199], Ti–N [200] and Ni–Al [201] alloys, as well as multiphase alloys [192,193] showed that only a 2- to 7-fold (for pure metals) and 2- to 4-fold (for multiphase alloys) increase in the yield stress τ_{y} is observed for real materials, instead of the 100-fold increase in the value of τ_{y} that is expected from the standard Hall–Petch law for materials with d as small as 10 nm. These experiments demonstrated that the value of τ_0 describing the lattice friction in the Hall–Petch relation can increase [238], and the value of the Hall–Petch coefficient K can decrease [238, 243] or even change sign [180, 182, 183, 192–194, 242, 245].

Ten years ago, there existed almost no theoretical models of the processes that could lead to the observed effects. Information on structure and properties of NCMs and grain boundaries was very inconsistent. It was actively debated as to whether one should treat the state of material in grain boundaries as a new state of matter which could not be considered as either crystalline or amorphous [168–172, 175, 177], because the state demonstrated even a total absence of the short-range translational order which is characteristic of the amorphous state. In such a situation, where authentic data on the structure and mechanisms of plastic deformation in NCMs were absent, it was decided to develop a generalized empirical analog of the Hall–Petch law which would allow for the yield characteristics of the materials of nanograins and intergranular layers, without dwelling on the particular physical processes that cause the yield stress in such materials [239].

Following Kocks [236], let us represent an NCM as a composite of a single crystalline matrix with inclusions of intercrystalline layers. The experimental value of the ratio of typical sizes of crystallites and interfaces is large enough to consider the interfaces in the matrix as thin plates which are chaotically placed and oriented. Since all possible orientations of the inclusions are equally probable, a medium with isotropic effective properties can be considered. Representing the inclusions by oblate ellipsoids, the effective shear modulus G of such a composite can be obtained [234] as

$$G = G_{\mathrm{m}} + c\frac{(G_{\mathrm{m}} + \eta G_{\mathrm{i}})(G_{\mathrm{i}} - G_{\mathrm{m}})}{G_{\mathrm{i}}(1+\eta)} , \qquad (2.3)$$

where G_{m} is the shear modulus of the matrix, G_{i} is the shear modulus of the inclusion (interface), $\eta = 0.5[1 + 3/(4 - 5\nu_{\mathrm{i}})]$, ν_{i} is Poisson's ratio of the inclusion, and c is the volume fraction occupied by inclusions ($c \ll 1$).

Equation (2.3) was obtained under the assumption that the matrix and the inclusions are elastic; no reservations were made about the ratio of their moduli [234]. It commonly occurs [246] that the yield stress of a crystalline material is in direct proportion to its shear modulus. Assume that the yield stress of the matrix, τ_{m}, of the inclusions, τ_{i}, and the effective yield stress of the composite, τ_{y}, are linearly connected with the corresponding shear moduli, i.e.,

$$G_{\mathrm{m}} = Q_{\mathrm{m}}\tau_{\mathrm{m}} , \qquad G_{\mathrm{i}} = Q_{\mathrm{i}}\tau_{\mathrm{i}} , \qquad G = Q\tau_{\mathrm{y}} , \qquad (2.4)$$

where Q, Q_{m} and Q_{i} are dimensionless factors. Additionally, assume that the values of τ_{m} and τ_{i} are connected with the typical structure period d corresponding to the grain size of the NCM by the following Hall–Petch-type relationships:

$$\tau_{\mathrm{m}} = \tau_{\mathrm{m}}^{*} + K_{\mathrm{m}}d^{-k/2} , \qquad \tau_{\mathrm{i}} = \tau_{\mathrm{i}}^{*} + K_{\mathrm{i}}d^{-l/2} , \qquad (2.5)$$

where k and l are integers. It is easily shown (e.g., for cube-shaped grains) that the volume fraction of the inclusions may be estimated as $c \approx 3\delta/d$, if the thickness δ is much less than the distance d between the interfaces. Then, substituting (2.4) and (2.5) into (2.3), the following empirical dependence of the effective yield stress of the composite on the typical structure size d is obtained [239]:

$$
\begin{aligned}
\tau_{\mathrm{y}} = & \frac{Q_{\mathrm{m}}}{Q}\left(\tau_{\mathrm{m}}^{*} + K_{\mathrm{m}} d^{-k/2}\right) + 3\frac{(1-\eta) Q_{\mathrm{m}} \tau_{\mathrm{m}}^{*} + \eta Q_{\mathrm{i}} \tau_{\mathrm{i}}^{*}}{(1+\eta) Q} \delta d^{-1} \\
& + 3\frac{1-\eta}{1+\eta} \frac{Q_{\mathrm{m}}}{Q} K_{\mathrm{m}} \delta d^{-(1+k/2)} + 3\frac{\eta}{1+\eta} \frac{Q_{\mathrm{i}}}{Q} K_{\mathrm{i}} \delta d^{-(1+l/2)} \\
& - 3\frac{\delta (Q_{\mathrm{m}})^2}{(1+\eta) Q Q_{\mathrm{i}}} \frac{(\tau_{\mathrm{m}}^{*})^2 + 2\tau_{\mathrm{m}}^{*} K_{\mathrm{m}} d^{-k/2} + (K_{\mathrm{m}})^2 d^{-k}}{\tau_{\mathrm{i}}^{*} d + K_{\mathrm{i}} d^{(1-l/2)}} . \qquad (2.6)
\end{aligned}
$$

The next step is to choose the values of k and l. In this case one must proceed from the specific physical mechanism that determines the yield stress in each of the phases in a particular situation. Assume that τ_{m} and τ_{i} are governed both by the stress in the heads of the dislocation pile-ups set against the matrix–inclusion boundary from the matrix side (Fig. 2.1a) and by the stress of the pile-ups in the immediate intergrain layer (Fig. 2.1b). The latter pile-ups provide a mechanism for intergrain sliding. The typical dimensions of pile-ups of both types are about d. This means that values of $k = l = 1$ may be chosen to get the usual Hall–Petch law for both the matrix and the inclusion.

Now consider how the quantities τ_{m}^{*}, τ_{i}^{*}, K_{m} and K_{i} are related to each other. If the main distinction of the interfaces from the crystallites is the low atomic density of the former, it seems reasonable to assume that $\tau_{\mathrm{i}}^{*} \approx q\tau_{\mathrm{m}}^{*}$ and $K_{\mathrm{i}} \approx pK_{\mathrm{m}}$, where q and p are dimensionless constants smaller than 1. Evidently, the lower the atomic density of the interface, the smaller the values of q and p.

Finally, for the sake of simplicity take $Q \approx Q_{\mathrm{i}} \approx Q_{\mathrm{m}}$. Then (2.6) is easily reduced to the form [239]

$$
\begin{aligned}
\tau_{\mathrm{y}} = & \tau_{\mathrm{m}}^{*} + K_{\mathrm{m}} d^{-1/2} - \frac{3}{1+\eta} \frac{\delta}{d}\left(q\tau_{\mathrm{m}}^{*} d + pK_{\mathrm{m}} d^{1/2}\right)^{-1} \\
& \times\left\{\left[1 - q(1-\eta+\eta q)\right](\tau_{\mathrm{m}}^{*})^2 d + \left[1 - p(1-\eta+\eta p)\right] K_{\mathrm{m}}^2 \right. \\
& \left. + \left[2 - q(1-\eta+\eta p) - p(1-\eta+\eta q)\right] \tau_{\mathrm{m}}^{*} K_{\mathrm{m}} d^{1/2}\right\} . \qquad (2.7)
\end{aligned}
$$

Consider in more detail the third term on the right-hand side of (2.7), which is accounted for by the fact that an interface is treated as an interlayer of finite thickness with its own elastic and plastic characteristics. Taking Poisson's ratio to lie in the interval 0.2–0.4, one obtains the following estimate for the value of η: $1 \leq \eta \leq 1.25$. For simplicity, we take $\eta \approx 1$. Hence, the atomic density of boundaries may amount to up to 70% [10,12,247]. At such densities the main mechanical properties of the material (the Young's modulus, yield stress, fatigue stress) are known to be 4–5 times worse than the characteristics of the compact body [248]. Therefore, one can estimate $q \approx p \approx 0.2$.

For these values of the parameters η, q and p, the third term in (2.7) turns out to be positive and is subtracted from the value of τ_{m}. As expected,

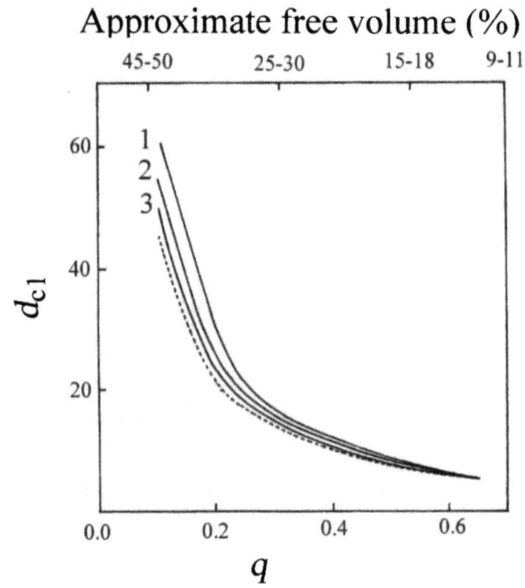

Fig. 2.2. Dependence of the critical grain size d_{c_1} on the parameter q for: (1) Ti, (2) Cu, (3) Fe, Ni and Ag [239]. The *dashed curve* is an approximation that may be applied to any material. Parameter q has been put in correspondence with the atomic density of boundaries using the data from [248]

the low-density interlayers lower the yield stress of the composite. Using the above estimates for η, (2.7) can be expressed in its simplest form:

$$\tau_y = \left(\tau_m^* + K_m d^{-1/2}\right)\left(1 - \frac{1-q^2}{q}\frac{3\delta}{2d}\right) . \tag{2.8}$$

Evidently, when such a dependence is plotted in $(\tau_y, d^{-1/2})$ coordinates, the straight line in the usual Hall–Petch law will remain straight only at large d or else as $q \to 1$. For $q < 1$, its slope will gradually decrease with decreasing grain dimension down to some critical size d_{c_1} where the effective Hall–Petch factor will change sign. The critical size d_{c_1} is easily found from the condition $\partial\tau_y/\partial(d^{-1/2}) = 0$. This equality can be reduced to the equation

$$(d_{c_1})^{-1/2} = -\frac{\tau_m^*}{3K_m} + \frac{1}{3}\sqrt{\frac{2q}{\delta(1-q^2)}}\sqrt{1 + \frac{1-q^2}{q}\frac{(\tau_m^*)^2\delta}{2K_m^2}} . \tag{2.9}$$

Substituting the usual values for parameters $\tau_m^*/K_m \approx 1$–$100\ \mathrm{mm}^{-1/2}$ and $\delta \approx 1$ nm, the following estimate for the critical grain size d_{c_1} is obtained [239]:

$$d_{c_1} \approx 9\delta\frac{1-q^2}{2q}\left[1 - 2\frac{\tau_m^*}{K_m}\sqrt{\frac{\delta(1-q^2)}{2q}}\right]^{-1} . \tag{2.10}$$

Figure 2.2 presents the family of curves $d_{c_1}(q)$ calculated according to (2.10) at $\delta = 1$ nm for Ti [249], Cu [242], Fe [250], Ni [237] and Ag [251]. The corresponding values of the ratio $\tau^*_{\mathrm{m}}/K_{\mathrm{m}}$ are 55.46, 26.26, 8.14, 4.35 and 3.91 $\mathrm{mm}^{-1/2}$, respectively. The dashed line is calculated according to the approximate formula $d_{\mathrm{c}_1} \approx 9\delta(1-q^2)/(2q)$ which may be used to estimate d_{c_1} in practically any material. At $q = 0.2$, this expression yields the grain size $d_{\mathrm{c}_1} \approx 22\delta = 22$ nm which corresponds to the volume fraction of boundaries $c \approx 0.12$.

Experimental values of d_{c_1} are spread over a wide range for different materials produced by different technologies. In [13,193,252–254], the values 8 [13], 11 [252–254], 25 [253, 254] and 45 nm [193] were reported. It is worth noting that the first value was observed in the electrodeposited nanocrystalline Ni–1.2%P alloys with very low porosity, while the others were attributed to multiphase materials produced by annealing of amorphous alloys. The authors of [13, 193] directly connected the microhardness decrease for small d with nanopores which were mainly distributed along grain boundaries and their triple junctions.

Equation (2.8) can be used to determine the other critical grain size, d_{c_2}, at which the yield stress of a composite becomes equal to the friction stress in the matrix (i.e., $\tau_{\mathrm{y}} = \tau^*_{\mathrm{m}}$). This critical size turns out to be equal to [239]

$$d_{\mathrm{c}_2} \approx \frac{3}{2}\frac{1-q^2}{q}\delta \approx \frac{d_{\mathrm{c}_1}}{3} \, . \tag{2.11}$$

For example, $\delta = 1$ nm and $q \approx 0.2$ give $d_{\mathrm{c}_2} \approx 7$ nm. Calculating the volume fraction of the interface phase using the more precise formula $c \approx 3d\delta/(d+\delta)^2$ (i.e., more precise than the former), we obtain $c \approx 0.33$ for the grain size.

Equation (2.8) turns out to agree pretty well with the different and apparently contradictory experimental data. For example, compare the results of the above calculations with the data from Chokshi et al. [242] and Jang and Koch [238]. Rewriting (2.8) in terms of the microhardness H and substituting the values of H^*_{m} and K_{m} corresponding to coarse-grained Cu [242] and Fe [238], the families of curves $H(d^{-1/2})$ are obtained for different values of q (see Fig. 2.3). The results given in [242] are seen to be in good agreement with the descending portion of the curve $H(d^{-1/2})$, with the difference from calculated values amounting to up to 25–45% at $q = 0.2$ (Fig. 2.3a). A qualitatively different increasing curve $H(d^{-1/2})$ [238] fits well between the calculated curves for $q = 0.8$ and $q = 0.9$ (Fig. 2.3b). The difference in dependences of H on $d^{-1/2}$ obtained in [242] and [238] seems to be caused by the fact that the measurements were carried out with samples in different states of grain boundaries [239]. The measurements in [242] were performed for a nanopowder-based, as-compacted, copper sample. The boundaries of the nanocrystallite grains in the sample thus produced correspond to the surface of the as-synthesized copper particles in the pre-compaction sample. Evidently, such a procedure for producing an NCM predetermines the low atomic density of boundaries (which increase as the internal stresses in the sample relax). The authors of [238] measured the microhardness of separate

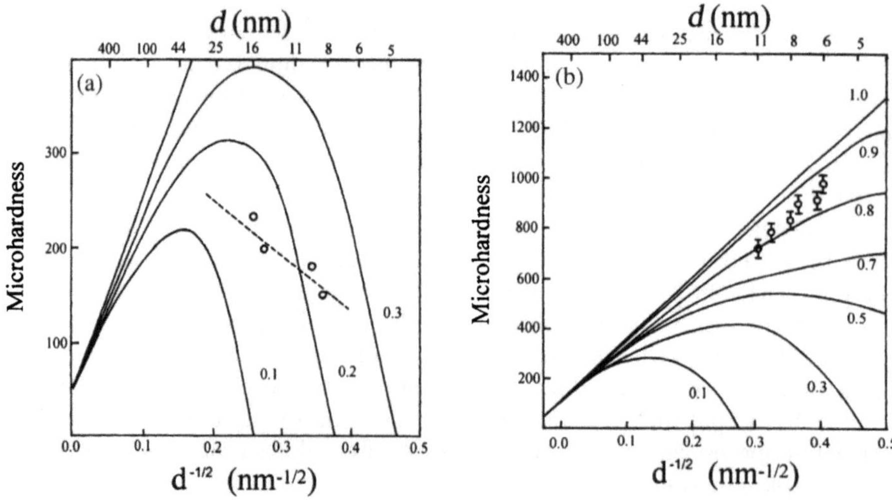

Fig. 2.3. Dependence of the microhardness on the grain size at different atomic densities of boundaries of nanocrystalline: (**a**) Cu, and (**b**) Fe [239]. This quantity is characterized by the parameter q. The numbers on the curves denote the values of this parameter. The *dashed straight line* was drawn through the experimental points from [242] (**a**) and [238] (**b**)

polycrystalline particles (with sizes ≤ 1 μm) consisting of nanocrystallites. In this case interfaces are formed from the walls of dislocation cells generated during the ball-milling process. It is clear that such a procedure produces nanocrystalline iron in which the atomic density of boundaries is significantly higher than in the former case. (According to Fig. 2.3b, it should differ from the normal value by about 5–7%.)

Therefore, the authors of [239] showed that dependences of τ_y and H on $d^{-1/2}$ for sufficiently small grains can differ from the Hall–Petch relation but still be consistent with experimental data. The contradictions arising on comparison of experimental results obtained by different authors can be resolved in the framework of the above approach. It was concluded [239] that the type of deviation from the Hall–Petch law is determined by the atomic density of the interfaces and the range of nanocrystallite sizes under study ($d < d_{c_1}$ or $d > d_{c_1}$).

Similar calculations based on the rule of mixture were carried out in [255–259]. The main differences were that, to describe the yield stress or microhardness of the intergranular phase, they chose the data for amorphous materials of close chemical composition [255–258] and accounted for the possibility that wedge disclinations may exist in nanocrystallites [255, 256] or at the triple junctions of grain boundaries [259]. Such models were supported by the following two arguments. First, studies of grain boundary structures in NCMs showed by the end of the 1980s [168–172, 175, 177] that the atoms in grain boundaries are in a totally disordered state. Therefore, they assumed that the characteristics of the intergranular phase would be close enough, as a first approximation, to the properties of the corresponding amorphous material. Second, at the same time, the first metallic glass–nanocrystal composites with Fe–Cr–B and TiNi alloy compositions [36–38] were synthesized in special regimes of high-pressure treatment [36, 37] and rapid cooling [38]. These composites consist of nanocrystallites divided by an amorphous interface, in which case the volume fractions of the nanocrystallite and amorphous phase are close. Concerning the disclination-like defects, many authors have discussed their role and behavior in the structure of amorphous and nanocrystalline materials (e.g., see [56, 260–262]) as well as in highly deformed polycrystalline metals [56, 263–266].

Let us consider a model [255, 256] where the simplest rule of mixture was used to estimate the yield stress of a metallic glass–nanocrystal composite, taking account of the influence of wedge disclinations. Let the model composite structure be nanocrystallites which are distributed uniformly in an amorphous matrix. Assume that some part of the nanocrystallites contains wedge disclinations. Under conditions of uniaxial tension, the simplest rule of mixture for the yield stress of such a composite then reads [255, 256]

$$\tau_{\rm y} = f\tau_{\rm am} + g\tau_{\rm cr} + h\tau_{\rm d} \ , \tag{2.12}$$

where $\tau_{\rm am}$ is the yield stress of amorphous interlayers, $\tau_{\rm cr}$ is the yield stress of a nanocrystallite without disclinations, $\tau_{\rm d}$ is the yield stress of a nanocrystallite with one or more disclinations, and f, g and h are the volume fractions of the above-mentioned components ($f + g + h = 1$).

Assume that $\tau_{\rm am}$ is constant and does not depend on the mean size of nanocrystallites while $\tau_{\rm cr}$ satisfies the Hall–Petch relationship: $\tau_{\rm cr} = \tau_{\rm cr}^0 + Kd^{-1/2}$, where $\tau_{\rm cr}^0$ is the lattice friction and K is the Hall–Petch coefficient. The term $\tau_{\rm d}$ is determined by the equations

$$\tau_{\rm d} = \begin{cases} \tau_{\rm cr} \ , & \text{if } \tau_{\rm cr} > \tau \ , \\ \tau \ , & \text{if } \tau_{\rm cr} < \tau \ . \end{cases} \tag{2.13}$$

Here a new parameter τ is introduced: the shear stress necessary for the motion of an edge dislocation intersecting an immobile wedge disclination. It is thus believed that the yield stress of a nanocrystallite with disclinations is either determined by its small size through the Hall–Petch law for $\tau_{\rm cr}$ or by the threshold stress τ [267], depending on which one of them, $\tau_{\rm cr}$ or τ, has the higher value. The latter may easily be estimated from the balance of forces acting on the dislocation after its intersection with the wedge disclination. This results in $\tau = |F|/(bd)$, where b is the Burgers vector of the dislocation and the force F is given by [267]

$$F(x) = -D\omega^2 b^2 \left\{ \frac{1}{2} + L\left[\frac{2l_1 + b}{l_1(l_1+b)} - \frac{1}{l_2}\right] - \frac{2\nu}{\omega}\ln\frac{L}{l_2} \right.$$
$$- \frac{\sqrt{l_1^2+\omega^2 b^2}}{4l_1} - \frac{\sqrt{(l_1+b)^2+\omega^2 b^2}}{4(l_1+b)}$$
$$\left. + \frac{1}{4}\ln\frac{\sqrt{l_1^2+\omega^2 b^2}+l_1}{\sqrt{(l_1+b)^2+\omega^2 b^2}-l_1-b} \right\} . \qquad (2.14)$$

Here x denotes the distance passed by the dislocation after it has intersected the disclination, $D = G/[2\pi(1-\nu)]$, ω is the disclination power, $2L$ the disclination length (in the case considered, this length is approximately equal to the nanocrystallite diameter, i.e., $2L \approx d$), $l_1 = x - \omega b$ and $l_2 = x + b$.

The force $F(x)$ reaches its maximum when $x = 2b$. In this event, $l_1 = (2-\omega)b$ and $l_2 = 3b$. Then, examining a dislocation–disclination intersection in a nanocrystallite with diameter $d \gg b$, one finds $L = d/2 \gg l_1, l_2$. In this situation the only term on the right-hand side of (2.14), which is proportional to L, provides the basic contribution to the maximum force $F(x = 2b)$. Taking only this contribution into account, the result for τ is [255, 256]:

$$\tau \approx \frac{D\omega^2}{6}\frac{9-\omega-\omega^2}{(2-\omega)(3-\omega)} . \qquad (2.15)$$

Let us estimate the value of the yield stress $\sigma_y = 2\tau_y$ for the metallic glass–nanocrystal Fe–Cr–B composite studied in [38]. Not knowing the appropriate numerical values for the given alloy, we used in [255, 256] the data taken for a pure iron. Hence, the shear modulus $G \approx 90$ GPa [268] allows one to estimate $\sigma_{am} = 2\tau_{am} \approx G/25 \approx 3.5$ GPa. The Hall–Petch parameters $\sigma_{cr}^0 = 2\tau_{cr}^0 \approx 66$ MPa [250] and $K' = 2K \approx 26447$ MPa nm$^{1/2}$ [250] lead for $d \approx 10$ nm to $\sigma_{cr} = 2\tau_{cr} \approx 8.3$ GPa. Substituting $G \approx 90$ GPa, $\nu \approx 0.3$ and $\omega \approx \pi/6$–$\pi/4$ into (2.15) results in $2\tau \approx G/21$–$G/8 \approx 4.3$–11.3 GPa. It then follows from (2.13) that $\sigma_d = 2\tau_d \approx 8.3$–$11.3$ GPa. Assuming that the nanocrystallites are very closely packed, as in [38], one can take the volume fractions as $f = 0.4$, $g = 0.5$ and $h = 0.1$. With all the above-mentioned values, (2.12) gives $\sigma_y = 2\tau_y \approx 6.7$ GPa. Here we represent the upper estimate of σ_y because the data for the real Fe–Cr–B alloy must be rather higher than those taken for a pure iron. Experiments [38] show that metallic glass–nanocrystal composites are specified by a low plasticity, in which case their yield stress σ_y is close to the strength $\sigma_b \approx 6$–6.5 GPa. Thus, the value obtained within the framework of the model [255, 256] is close to that revealed experimentally in [38].

One can conclude that the model of metallic glass–nanocrystal composites [255, 256] which allows for the presence of disclinations in a part of the nanocrystallites can explain the unusually high mechanical properties (the microhardness, yield stress and strength) of such materials. The main reasons for the significant increase in these properties compared with typical homogeneous metallic glasses are the high properties of nanocrystallites themselves due to their small dimensions and the role of disclinations as obstacles for dislocation motion in nanocrystallites, combined with the very high concentration of nanocrystallites in the amorphous matrix.

A very similar model which did not take into account the role of disclinations was proposed later by Carsley et al. [257] to describe the microhardness of NCMs. They wrote the rule of mixture (2.12) at $h = 0$ in terms of the microhardness as [257]

$$H = \frac{(d-\delta)^3}{d^3}\left(H_0 + \beta d^{-1/2}\right) + \frac{3d^2\delta - 3d\delta^2 + \delta^3}{d^3}\frac{\mu_{\mathrm{cr}}}{12}, \tag{2.16}$$

where H is the microhardness of the NCM as a whole, H_0 and β are the appropriate Hall–Petch constants of a nanocrystallite, and μ_{cr} is its shear modulus. To obtain (2.16), the relations $H_0 = 3\sigma_{f0}(\varepsilon = 8\%)$ and $\beta = 3\beta'(\varepsilon = 8\%)$ were taken from [269] and used to connect the constants H_0 and β with the corresponding parameters for the yield stress. As the properties of the intergranular phase were treated as those of the appropriate metallic glass, its microhardness H_{gb} was supposed to be constant and approximately equal to 1/6 the shear modulus μ_{gl} of the metallic glass [270–272], while the latter was approximated as half that of the crystalline metal, $\mu_{\mathrm{cr}}/2$.

Equation (2.16) was analysed in [257] with experimental values for parameters attributed to nickel [240,273], iron [274] and copper [275]. The thickness of grain boundaries was chosen as 1 nm. The $H(d^{-1/2})$ plots demonstrate linear portions in the field $d > 25$ nm with subsequent deviations from linearity for $d < 25$ nm. Near the point $d \approx 5$ nm, the curves achieve maximum values and decrease further at $d < 5$ nm, approaching the constant level of the microhardness calculated for the metallic glass. Comparison with the experimental data for nickel [240, 273], iron [274] and copper [275] showed that the model by Carsley et al. [257] correctly predicts the general shape of the $H(d^{-1/2})$ curves for all three metals, and correlates well the measurements with nickel and iron. In these cases, the experimental points are close to the rising branches of the curves. For copper, however, the curve goes above the experimental points and reaches a magnitude of about 4 GPa that is 1.5–2 times higher than the maximal measured values. Moreover, the position of the maximum of the theoretical curve ($d \approx 5$ nm) as well as the position of its dropping branch ($d < 5$ nm) turn out to be severely shifted towards the field of small grain sizes, in contrast to the experimental points [242, 276]. (Observations [13, 193, 252–254] carried out later have also testified to this mismatch.) In the authors' opinion [257], a possible reason for this discrepancy is that they did not take into account the residual porosity of nanocrystalline copper. Let us remember that this point was treated as the key factor in the model by Gryaznov et al. [239] when considering the differences between the $\tau_{\mathrm{y}}(d^{-1/2})$ curves for iron [238] and copper [242].

One more similar model was proposed by Kim [258]. As in [255–257], the properties of the intergranular phase were described as those of the metallic glass. However, in discussing the properties of nanocrystallites, Kim [258] used the conclusion by Wang et al. [277] about the existence of a critical nanocrystallite size d_c at which the usual dislocation mechanisms of plasticity and strengthening stop working. Accordingly, they believe that the Hall–Petch relation is true only for the case $d \geq d_c$, while in smaller grains ($d \leq d_c$), there is no strengthening effect and the regime of ideal plasticity is realized when the microhardness is constant and does not depend on the grain size. The numerical estimates of the critical size d_c made by Wang et al. [277] and represented in [258] are rather spread out as 1.2 nm $< d_{c,Al} <$ 59.3 nm, 1.7 nm $< d_{c,Cu} <$ 39.4 nm, 1.3 nm $< d_{c,Ni} <$ 25.6 nm and 1.8 nm $< d_{c,Pd} <$ 75.3 nm.

In agreement with the above assumptions, the rule of mixture for the microhardness of an NCM was chosen as [258]

$$H = \begin{cases} (H_0 + \beta d^{-1/2}) f_{cr} + H_{ic}(1 - f_{cr}) \,, & d \geq d_c \,, \\ (H_0 + \beta d_c^{-1/2}) f_{cr} + H_{ic}(1 - f_{cr}) \,, & d \leq d_c \,, \end{cases} \tag{2.17}$$

where H_{ic} is the microhardness of the intergranular phase and f_{cr} is the volume fraction of nanocrystallites. The calculations were carried out with the same three metals (Ni, Fe and Cu) as in [257] and the same parameter values. The critical grain sizes taken for nickel and copper were as above and the estimate 3.4 nm was chosen for iron [278]. It is rather natural that, using the low limits of the ranges for d_c, Kim [258] obtained practically the same $H(d^{-1/2})$ curves as the authors of [257], with characteristic maximums at $d = d_{peak} \approx$ 3–4 nm. The maximum condition gives [258]

$$d_{peak} = \left[\frac{3\delta}{\beta}(H_0 - H_{ic}) + \sqrt{\left[\frac{3\delta}{\beta}(H_0 - H_{ic})\right]^2 + 7\delta} \right]^2 \,, \tag{2.18}$$

and the numerical estimates are $d_{peak} \approx$ 2.9 nm (Cu), 3.9 nm (Ni), and 4.3 nm (Fe). Until $d_c < d_{peak}$, the shape of the $H(d^{-1/2})$ curve is conserved with both the maximum point and dropping branch (for $d < d_{peak}$). However, if one takes $d_c > d_{peak}$, the curve shape changes drastically in the range $d \leq d_c$ where the curve has a break (the slope decreases with a jump) at the point $d = d_c$ and then rises monotonically with decreasing grain size. The model [258] thus predicts the two qualitatively different shapes of the $H(d^{-1/2})$ curve depending on the value of the critical grain size. However, both of them demonstrate significant deviations from the usual Hall–Petch law in the field of nanoscopic grain sizes.

The theoretical $H(d^{-1/2})$ curves of [258] agree well with the experimental data for Ni [240,273] and Fe [238], as in the case of [257]. For copper [242,279–281], however, the experimental points of [281] are only close to the curve.

All other experimental points lie much lower than the calculated curve. Like the authors of [257], Kim [258] explained the reasons for the discrepancy by the need to account for the residual porosity of NCMs.

On the basis of the results reported in papers [255–258], one can conclude that the description of the properties of the intergranular phase through the properties of an appropriate metallic glass seems to be correct only in those cases when one considers either truly metallic glass–nanocrystal composites [255, 256] or NCMs with rather large (larger than ≈ 20 nm) grain sizes [257, 258]. Use of this approach in models of the NCMs with small (near or smaller than ≈ 20 nm) grain sizes results in overestimates of the yield stress or microhardness and in a shift of their maximal values to the field of very small (near or smaller than ≈ 5 nm) grain sizes [257, 258]. The authors of [259, 282] either withdrew the idea of using the properties of metallic glass to model the intergranular phase [259] or else, instead of taking a constant value for the yield stress of the metallic glass as in [255–258], used a variable quantity [282]. (More precisely, it was taken constant in the region of large grain size but then sharply decreasing in proportion with the decreasing grain size.) Let us consider these later results.

Konstantinidis and Aifantis [259] used a rule of mixture which is completely similar to (2.16) or to the first line of (2.17). The only difference was the assumption that the microhardness of the intergranular phase also satisfies the Hall–Petch relation with its effective parameters $H_{0\mathrm{GB}}$ and β_{GB}. (This is, in fact, the limiting case of the model by Gryaznov et al. [239].) For simplicity, the authors [259] assumed that $H_{0\mathrm{GB}} = H_0$, i.e., the friction stress was assumed to be equal for both the intergranular and crystallite phases. In its turn, β_{GB} was estimated as the Hall–Petch β coefficient multiplied by a factor which would allow for the presence of obstacles for dislocation glide within grain boundaries. As obstacles, triple junctions of grain boundaries were considered, including those containing triple line disclinations. Within the approximation of line tension for dislocations, the following estimate was obtained [259]:

$$\beta_{\mathrm{GB}} = \beta \ln \frac{\vartheta d}{r_0} \Big/ \ln \frac{\vartheta d'_{\mathrm{c}}}{r_0} \,, \tag{2.19}$$

where ϑ is a numerical factor less than one (the ratio of the radius of curvature of a slightly curved dislocation line to the grain size), r_0 is the radius of the dislocation core, d'_{c} is the critical grain size, such that for $d < d'_{\mathrm{c}}$ the dislocation glide is no longer prohibited by the line tension and dislocations bypass obstacles by Orowan's mechanism [246]. The final expression for the microhardness of an NCM then reads [259]

$$H = H_0 + \beta \left[\frac{(d-\delta)^3}{d^3} + \frac{d^3 - (d-\delta)^3}{d^3} \ln \frac{\vartheta d}{r_0} \Big/ \ln \frac{\vartheta d'_{\mathrm{c}}}{r_0} \right] d^{-1/2} \,. \tag{2.20}$$

The authors [259] carried out a detailed comparison of their $H(d^{-1/2})$ curves corresponding to (2.20) with the data of many experimental measurements which were available at that time. The curves were drawn for a number of nanocrystalline metals (Cu, Fe, Pd, Ti, Ni) and intermetallics (NiP, NiZr, TiAl, Nb_3Sn, FeMoSiB, FeCoSiB, FeSiB, PdCuSi). The fitting parameter ϑ, which varied from 0.0065 (for FeMoSiB) to 0.4 (for Fe), was used to fit the theoretical curves to the experimental points. In all cases, the agreement is quite good. As a result, the calculated values of the microhardness achieve their maximum at the critical grain size d'_c which varies with different materials from 6 nm (for Fe) to 50 nm (for FeMoSiB). It was shown that the microhardness of several metals (Cu, Pd) and of all the intermetallics under investigation starts to drop with further decrease in the grain size.

Thus, significant deviations of the microhardness from the Hall–Petch law and its decrease in the region of small grain sizes is explained in [259] by peculiarities of the interaction of dislocations gliding along the grain boundaries with the obstacles located there. The critical grain size in the model [259] is the one at which the interaction mechanism changes from dislocation pinning to Orowan bypassing. The maximal microhardness of an NCM is then achieved at the critical grain size.

The rule-of-mixture approach has also been used in models [277, 283] of the abnormal Hall–Petch relationship in NCMs. Wang et al. [277] have elaborated a model describing an NCM as a composite with crystallites, grain boundaries, their triple junctions and quadruple nodes playing the role of constituent phases. The model [277] led to a third order polynomial for the grain size dependence of the yield stress, which accounts for experimental data. Ovid'ko [283] has demonstrated that grain boundaries of nanoscale length are more often quasiperiodic than periodic (see also [284]), in which case they exhibit enhanced plastic properties. Following the model [283], quasiperiodic tilt boundaries are capable of effectively contributing to the experimentally observed abnormal Hall–Petch relationship in NCMs.

A further and very substantial development of the rule-of-mixture approach was given by Kim et al. [282]. In describing the properties of nanocrystallites, they introduced the evolution of dislocation density and diffusion creep, while the model of diffusion flow was considered as the mechanism of grain boundary plasticity. The rule of mixture was written in the simplest form:

$$\sigma = \sigma_{cr} f_{cr} + \sigma_{ic}(1 - f_{cr}) , \qquad (2.21)$$

where σ_{cr}, σ_{ic}, and σ denote the stress in nanocrystallites, the stress in grain boundaries, and the overall stress in the NCM, respectively. σ_{cr} and σ_{ic} were calculated using the assumption that the strains in both phases are the same and equal to the macroscopic applied strain.

In a crystallite, three mechanisms of plastic deformation were considered [282] to contribute to the total deformation behavior:

- the dislocation glide mechanism,
- the lattice diffusion mechanism, in which vacancies diffuse through the bulk of the crystallite,

- the boundary diffusion mechanism, in which vacancies diffuse along crystallite boundaries.

These mechanisms were postulated to contribute simultaneously to the total crystallite deformation. For monotonic uniaxial loading, the total strain rate was assumed to be separable into elastic and plastic components. The total plastic strain rate $\dot{\varepsilon}^{\mathrm{p}}$ was written as a superposition of contributions from the individual mechanisms listed above [282]:

$$\dot{\varepsilon}^{\mathrm{p}} = \dot{\varepsilon}^{\mathrm{p}}_{\mathrm{c,u}} + \dot{\varepsilon}^{\mathrm{p}}_{\mathrm{c,b}} + \dot{\varepsilon}^{\mathrm{p}}_{\mathrm{c,l}} \,, \tag{2.22}$$

where the subscript c refers to the crystallite phase and u, b and l stand for the deformation mechanisms, i.e., the dislocation glide described within a dislocation-density-related unified constitutive model [285], boundary diffusion mechanism and lattice diffusion mechanism, respectively.

The first term in (2.22) is determined through the following system of three equations [282, 285]:

$$\dot{\varepsilon}^{\mathrm{p}}_{\mathrm{c,u}} = \dot{\varepsilon}_*(\sigma/\sigma_0)^m Z^{-m/2} \,, \tag{2.23}$$

$$\frac{\mathrm{d}Z}{\mathrm{d}\varepsilon^{\mathrm{p}}_{\mathrm{c,u}}} = C + C_1\sqrt{Z} - C_2 Z \,, \tag{2.24}$$

$$C_2 = C_{20}\left(\frac{\dot{\varepsilon}^{\mathrm{p}}_{\mathrm{c,u}}}{\dot{\varepsilon}_0}\right)^{-1/n} . \tag{2.25}$$

Equation (2.23) gives a relation between the equivalent (von Mises) plastic strain rate $\dot{\varepsilon}^{\mathrm{p}}_{\mathrm{c,u}}$ associated with the unified constitutive model of a crystallite and the equivalent stress σ. Here Z is the dislocation density normalized by its initial value. The quantity σ_0 is related to the initial dislocation density. The exponent m is stress independent for a face-centred cubic (f.c.c.) metal. The parameter $\dot{\varepsilon}_*$ is a constant normalization factor with units of strain rate. Equations (2.24) and (2.25) describe the evolution of the dislocation density with the equivalent (von Mises) plastic strain rate $\dot{\varepsilon}^{\mathrm{p}}_{\mathrm{c,u}}$. The parameters C, C_1, C_{20}, σ_0, $\dot{\varepsilon}_*$, and $\dot{\varepsilon}_0$ were taken to be constant for f.c.c. metals at low temperatures. The exponents m and n are inversely proportional to the absolute temperature T and were also taken as constant for a given temperature T.

The plastic strain rate of a crystallite associated with the lattice diffusion mechanism (Nabarro–Herring creep) is given by

$$\dot{\varepsilon}^{\mathrm{p}}_{\mathrm{c,l}} = 14\frac{\Omega\sigma}{kT}\frac{D^{\mathrm{sd}}_{\mathrm{ld,0}}}{d^2}\exp\left(-\frac{Q_{\mathrm{ld}}}{kT}\right) , \tag{2.26}$$

and that associated with the grain-boundary diffusion mechanism (Coble creep) by

$$\dot{\varepsilon}^{\mathrm{p}}_{\mathrm{c,b}} = 14\pi\frac{\Omega\sigma}{kT}\frac{\delta D^{\mathrm{sd}}_{\mathrm{bd,0}}}{d^3}\exp\left(-\frac{Q_{\mathrm{bd}}}{kT}\right) . \tag{2.27}$$

Here k is the Boltzmann constant, Ω is the atomic volume, $Q_{\rm ld}$ is the activation energy for lattice self-diffusion, and $Q_{\rm bd}$ is the activation energy for boundary self-diffusion. The pre-exponential factors $D^{\rm sd}_{\rm ld,0}$ and $D^{\rm sd}_{\rm bd,0}$ refer to lattice diffusion and grain-boundary diffusion, respectively.

It is worth noting that in [282] (as before in [258]), a critical grain size $d_{\rm c}$ ($\approx$ 8.2 nm for Cu [277]) was introduced, such that the dislocation glide mechanism of plastic deformation works in large enough nanocrystallites with $d > d_{\rm c}$. In smaller nanocrystallites, only the diffusion mechanisms were considered to be active.

In describing the deformation behavior of the intergranular phase, the authors [282] believed that the plastic deformation was associated with diffusional mass transport along the boundaries at small grain sizes, while at large grain sizes, the grain-boundary interlayers behaved like a metallic glass under homogeneous deformation. Accordingly, the strength of the intergranular phase was assumed to increase up to an upper limit and then to stay at that level as the grain size increased. This upper limit was set to be the strength of the material in the amorphous state. The minimal grain size which corresponds to this upper limit as well as the values of the yield stress were determined by the strain rate of the intergranular phase [282]:

$$\dot{\varepsilon}_{\rm bd} = \frac{\Omega_{\rm b}\sigma}{kT}\frac{(2d-\delta)D^{\rm sd}_{\rm bd}}{d^3}\,, \qquad (2.28)$$

where $\Omega_{\rm b}$ is the atomic volume in the grain boundary, σ is the component of the local stress normal to the boundary, and $D^{\rm sd}_{\rm bd}$ is the coefficient of self-diffusion within the grain boundary.

It was noted in particular [282] that in the model under consideration, the same grain boundaries were regarded as 'channels' for material flow of two phases: the boundary phase itself and the crystallite. In other words, two separate flows coexist in one such channel. One of the flows provides the boundary diffusion in crystallites and the Coble creep mechanism, while the other flow is responsible for plastic deformation of the grain boundaries.

The model of [282] was analysed with the data for nanocrystalline copper. The yield stress was determined for the crystallite phase by substituting (2.22) with (2.23)–(2.27) into the usual governing equation for the theory of plasticity, and using the expression (2.28) for the intergranular phase. The overall yield stress of the NCM was calculated with the rule of mixture (2.21). The stress–strain curves and the 0.2% offset stress $\sigma_{0.2}$ for the crystallite phase were studied in wide ranges of grain sizes (from 10 nm to 1000 nm) and strain rates (from 10^{-5} s^{-1} to 10^{-3} s^{-1}).

As the grain size diminished, the stress $\sigma_{0.2}$ first increased, achieved its maximal value, and then decreased. The deviations of the $\sigma_{0.2}(d^{-1/2})$ curve from the usual linear Hall–Petch law began earlier as the strain rate was smaller. Accordingly, the maximal value of the stress $\sigma_{0.2}$ was smaller and was reached earlier (from $\approx$ 300 MPa at $d \approx$ 100 nm and $\dot{\varepsilon} = 10^{-5}$ s^{-1} to $\approx$ 600 MPa at $d \approx$ 20 nm and $\dot{\varepsilon} = 10^{-3}$ s^{-1}).

The investigation of the relative contributions of different mechanisms (dislocation glide, and boundary and lattice diffusion) to the strain rate of nanocrystallites showed that the contribution of lattice diffusion may be neglected in any practical case. At small grain sizes, boundary diffusion is the dominant deformation mechanism, while in coarse grains, the strain rate for the dislocation glide mechanism dominates completely. The relative contributions of the dislocation glide and boundary diffusion mechanisms become equal at approximately 16 nm (for $\dot{\varepsilon} = 10^{-3}$ s^{-1}) and 55 nm (for $\dot{\varepsilon} = 10^{-5}$ s^{-1}). In accordance with this analysis, the $\sigma(\varepsilon)$ curves demonstrated typical deformation strengthening of crystallites with sizes 100 nm and 1000 nm, where the contribution of the dislocation glide mechanism still dominates, and the complete absence of the strengthening in nanocrystallites with sizes of 10 nm, where Coble creep by boundary diffusion prevails.

A comparison of the overall $\sigma(\varepsilon)$ and $\sigma_{0.2}(d^{-1/2})$ curves [282] with the results of experiments [233, 244, 281, 286, 287] carried out with samples of poly- and nanocrystalline copper having low porosity showed, in general, a good agreement, although the model [282] sometimes gives quite overestimated or underestimated values. A great achievement is that the maxima of the $\sigma_{0.2}(d^{-1/2})$ curves shift to the range of coarser grains (now from ≈ 16 nm at $\dot{\varepsilon} = 10^{-3}$ s^{-1} to ≈ 70 nm at $\dot{\varepsilon} = 10^{-5}$ s^{-1}) in contrast to the previous models [257, 258], which gave values near or smaller than 5 nm. This is a natural corollary of the refusal to consider the intergranular phase as an amorphous material, in the range of small grain sizes. The remaining discrepancies between the predictions of the model and the experimental data were referred by the authors [282] to the experimental uncertainties and also to the need to take into account the influence of the residual porosity of NCMs. Note that the last reason was already mentioned in earlier works [257, 258].

The authors [282] thus took an essential step forward in modeling the plastic deformation of NCMs within the rule-of-mixture approach. Allowing different possible mechanisms of crystallite plasticity and withdrawing the representation of grain boundaries through the interlayers of an amorphous material with a constantly high value of strength, they obtained quite realistic curves $\sigma(\varepsilon)$ and $\sigma_{0.2}(d^{-1/2})$ over a wide range of strain rates. It is natural that such a refined model had to be restricted in some other respects (only uniaxial loading, only f.c.c. metals, only low porosity). However, the authors [282] claimed that a new quantitative model was in progress which would incorporate the effects of porosity and grain size distribution.

Concluding the discussion of the aforementioned models which aimed to obtain, within the rule-of-mixture approach, the Hall–Petch-type dependences for NCMs and to study the reasons and peculiarities of the deviations from the classical linear law $\sigma_y(d^{-1/2}) \sim d^{-1/2}$, one can make the following inferences:

- The rule-of-mixture approach is a simple and effective way to obtain and study the Hall–Petch-type relation for NCMs.

- The rule of mixture may be used in a general empirical form without dwelling on the particular physical mechanisms of plastic deformation of crystallites and grain boundaries, just as it can account for different mechanisms of motion and interaction of defects in both (crystallite and intergranular) phases.
- Models of the intergranular phase based on the similarity with the interlayers of an amorphous material having a constantly high value of strength seem to be applicable only to the real metallic glass–nanocrystal composites and inapplicable to conventional NCMs where these models lead to essential disagreements with experimental data in the range of small grain sizes.
- One of the key problems in describing the Hall–Petch-type relation for real NCMs is to account for the residual porosity of samples.

2.2 Key Mechanisms of Plastic Deformation

In this section we will consider theoretical models directly focusing on physical mechanisms of plastic flow in NCMs. In doing so, we will first discuss models based on the idea of conventional lattice dislocation motion in nanograin interiors as the dominant deformation mechanism in NCMs (see Sect. 2.2.1). We then consider deformation mechanisms associated with the active role of grain boundaries in NCMs (see Sect. 2.2.2). We also analyse the combined effects of the competition between various deformation mechanisms (depending on the grain size) and a distribution of grain size on the deformation behavior of NCMs (see Sect. 2.2.3). Finally, we consider the role of enhanced triple junction diffusion in plastic deformation processes in NCMs (see Sect. 2.2.4).

2.2.1 Lattice Dislocation Motion

There are models that treat motion of lattice dislocations in grain interiors as the dominant deformation mechanism in NCMs, as with coarse-grained polycrystals. Within this framework, the experimentally documented deviations from the conventional Hall–Petch relationship in NCMs are explained as being related to the influence of grain size reduction and high-density ensembles of grain boundaries on the formation of lattice dislocation pile-ups in grain interiors and the penetration of lattice dislocations through grain boundaries in such materials.

We begin with a brief discussion of the models describing the classical Hall–Petch relationship (2.1) in coarse-grained polycrystals. Most of these models can be rationalized in terms of a dislocation pile-up model. These are reviewed in detail by Li and Chou [288]. In deriving the Hall–Petch relation, the role of grain boundaries as a barrier to dislocation motion is considered in various models. In one type of models [289–291], a grain boundary acts as a barrier causing dislocations to pile up and stresses to concentrate, thereby activating dislocation sources in neighboring grains and initiating the slip from grain to grain. In another type of models [232, 292, 293], grain boundaries are regarded as dislocation barriers limiting the mean free path of the dislocations and thereby increasing strain hardening, which results in a Hall–Petch-type relation. The various competing theories of strengthening by grain refinement have been discussed by several authors. For a recent survey, see Lasalmonie and Strudel [237]. It is clear that a variety of processes, both dislocation- and non-dislocation-based, could be postulated. It is possible that several of these processes could compete with one another or reinforce the deformation process.

We now turn to a discussion of the models focusing on the lattice dislocation mechanism of plastic flow in NCMs. First, let us consider the dislocation model of Pande and Masumura [294]. The assumption made in this model is that the classical Hall–Petch dislocation pile-up model is still dominant except that the analysis must take into account the fact that, in nanometer size grains, the number of dislocations within a grain cannot be very large. Further, at still smaller grain sizes, this mechanism should cease when there are only two dislocations in the pile-up. Pande and Masumura [294] showed that a dislocation theory for the Hall–Petch effect gives a linear dependence of τ on $d^{-1/2}$ only when there is a large number of dislocations in a pile-up and plasticity is not source-limited. In this regime, the yield stress increases as d decreases because the pile-ups contain fewer dislocations, the stress concentration at the head falls, and a larger applied stress is required to compensate. When the number of dislocations falls to one, no further increase in the yield stress is possible and it saturates. If, however, sources must operate in each grain, an additional component of the yield stress exists, viz., at least Gb/d, where G is the shear modulus and b is the magnitude of the Burgers vector. The yield stress should then rise as d^{-1}. Thus from these arguments, at a small grain size, either the yield stress should rise faster than $d^{-1/2}$ or it should saturate, but it should not decrease. If the number of dislocations n in a pile-up is not too large, the length of the pile-up L is not linear in n, but an additional term is necessary. Chou [295] gives the relation

$$L \simeq \frac{A}{2\tau}\left|4\left[n+m-1-2i_1\left(\frac{2n}{3}\right)^{1/3}\right]\right| , \tag{2.29}$$

where $A = Gb/[2\pi(1-\nu)](= Gb/(2\pi))$ for edge (screw) dislocations and $i_1/6^{1/3} = 1.85575$. Pande and Masumura [294] give the improved expression

$$L \simeq \frac{A}{2\tau}\left[2(n+m-1)^{1/2} - \frac{i_1+\varepsilon}{12^{1/3}(n+m-1)^{1/6}}i_1\right] , \tag{2.30}$$

where ε is a small correction term ($\varepsilon \ll 1$) which can be neglected. They find that, for small grain sizes, there are additional terms in the Hall–Petch relation, viz.,

$$s = \lambda^{-1/2} + c_1 \left(\lambda^{-1/2}\right)^{5/3} + c_2 \left(\lambda^{-1/2}\right)^{7/3} , \qquad (2.31)$$

where $s = \tau/(m\tau^*)$, $c_1 = -0.6881$, $c_2 = 0.21339$ and $\lambda = Lm\tau^*/2A$. This model recovers the classical Hall–Petch relation at large grain sizes, but for smaller grain sizes, τ levels off. This model cannot therefore explain a drop in τ.

Valiev et al. [296] mention a creep mechanism in which dislocations require a specific amount of time to be absorbed into the grain boundary. Such times involve a characteristic length of 30 nm, which is approximately the distance at which the image of the dislocation 'disappears'. But a detailed mechanism relating τ to d in this region is not available.

In a recent publication, Malygin [297] seems to have provided just such a theory based on a dislocation mechanism. The dislocation density $\rho(d)$ at any grain size d is related in the usual fashion to the square of the yield stress and an expression is obtained that connects ρ to the grain size. The expression is based on the assumption that grain boundaries in NCMs act predominately as sinks for lattice dislocations (just the opposite to that used by Li [298], who postulated that grain boundaries could be sources for dislocation generation). In Malygin's model, as the grains become finer and finer, more and more dislocations are absorbed by the grain boundaries, leading ultimately to a drop in dislocation density and hence in a flow stress, since the two are directly related, as mentioned above. The model is attractive and should be considered further. At present, we merely point out two problems with it. First, it is doubtful that the dislocations play the same role whether the grains are large or small. It is more likely that dislocations in ultrafine grains, if present at all, are confined to grain boundaries [293]. Second, in Malygin's model [297], the stress calculated is a work-hardened flow stress rather than a yield stress.

Lu and Sui [299] assume that both the energy and free volume of grain boundaries decrease with a reduction in the grain size d. This gives rise to an enhancement of lattice dislocation penetration through grain boundaries and a corresponding softening of NCMs. Following Scattergood and Koch [300], the yield stress of fine-grained materials is controlled by the intersection of mobile lattice dislocations with the dislocation networks at grain boundaries. In this context, assuming that the dislocation line tension is size dependent, there is a critical grain size that corresponds to a transition from cutting to Orowan bypassing of the dislocation network. This critical grain size characterizes the experimentally detected transition from the conventional to inverse Hall–Petch relationship with falling grain size d. Nazarov et al. [301], Nazarov [302] and Lian at al. [303] have developed models similar to that of Pande and Masumura [294], focusing on the influence of the grain size d on the parameters of lattice dislocation pile-ups in grain interiors.

Zaichenko and Glezer [304] have described the role of grain boundary disclinations – rotational defects formed at triple junctions – as sinks and sources of the lattice dislocations moving in grain interiors and causing the plastic flow in NCMs. Following the model [304], lattice dislocations are emitted and absorbed at the opposite triple junctions of grain boundaries, containing disclinations (formed during fabrication of NCMs). In doing so, the yield stress is caused by the interaction between grain boundary disclinations and lattice dislocations moving in grain interiors. Disclinations move in the vicinity of their initial positions (corresponding to the non-deformed state of the material) at the grain boundary junctions when they emit and absorb the lattice dislocations.

Despite the good correspondence between theoretically predicted $\tau(d)$ dependences and experimental data, all the above models based on representations of the lattice dislocation mechanism of plastic flow in NCMs face the question of whether the lattice dislocations exist and play the same role in nanograin interiors as with conventional coarse grains. As pointed out in papers [15, 56, 59, 77, 78, 86, 305], the existence of dislocations either in free nanoparticles or nanograins composing a nanocrystalline aggregate is energetically unfavourable if their characteristic size, nanoparticle diameter or grain size, is lower than some critical size which depends on such material characteristics as the shear modulus and the resistance to dislocation motion. The dislocation instability in nanovolumes is related to the effect of the so-called image forces occurring due to the elastic interaction between dislocations and either the free surface of the nanoparticle or grain boundaries adjacent to a nanograin. The paucity of mobile dislocations in nanograins has been well documented in electron microscopy experiments [189, 196, 198, 224, 306].

2.2.2 Evolution of Grain Boundary Defect Structures

Let us discuss theoretical models which describe the plastic flow in NCMs as occurring mostly via deformation mechanisms associated with the active role of grain boundaries. Such deformation mechanisms are, in particular, grain boundary sliding and diffusion plasticity mechanisms related to grain boundary diffusion.

Hahn et al. [307] treat grain boundary sliding as a deformation mechanism effectively competing with the conventional lattice dislocation motion in NCMs. In doing so, grain boundary sliding occurs via the motion of mobile grain boundary dislocations with the Burgers vectors parallel to the boundary planes. Triple junctions of grain boundaries, where boundary planes with various orientations join together, serve as obstacles for grain boundary dislocation motion. In order to overcome the obstacles, grain boundary structures undergo grain boundary sliding accommodation transformations; the same comes into play in the superplastic deformation characterized by dominant grain boundary sliding. The model of Hahn et al. [307] describes a local migration of grain boundaries as the accommodation mechanism for grain boundary sliding. More precisely, the local migration of grain boundaries is considered to form an approximately planar ensemble of grain boundaries along which an intensive plastic shear occurs via the correlated grain boundary sliding processes. These macroscopic planar grain boundary structures are associated with the shear bands where high (super)plastic deformation is localized, resulting in a large macroscopic deformation of a nanocrystalline sample.

The local migration of grain boundaries occurs more effectively than other diffusion accommodation mechanisms, because the characteristic spatial scale of diffusion associated with the local migration of grain boundaries is very small, being of the order of the grain boundary thickness. In these circumstances, the strain-rate-controlling process is grain boundary sliding. Although there is experimental evidence for the local migration of grain boundaries in superplastically deformed polycrystals [308], there is at present no direct experimental data confirming the model of Hahn et al. [307] in the case of NCMs. Furthermore, this model does not quantitatively describe the yield dependence on the grain size.

As has been noted above, some permanent transformations of grain boundary defect structures are required to save the continuity of a polycrystalline material. Several types of transformation of grain boundary dislocation pile-ups at triple junctions of grain boundaries (Fig. 2.4) were considered in detail in [309]. Let an initial pile-up of n edge dislocations with Burgers vectors $\boldsymbol{b}$ be generated under an external shear stress τ (Fig. 2.4a). The growing value of τ may stimulate the following transformations in the defect structure near the triple junction.

(I) The 'head' dislocation of the pile-up (with Burgers vector $\boldsymbol{b}$) splits into two new grain boundary dislocations (with Burgers vectors $\boldsymbol{b}_1$ and $\boldsymbol{b}_2$, respectively) that move along the adjacent grain boundaries (Fig. 2.4b). This process may occur repeatedly, even to the extent of splitting all the dislocations in the pile-up (Fig. 2.4c).

(II) The head dislocation of the pile-up splits into an immobile dislocation with Burgers vector $\boldsymbol{b}_1$, which stays at the triple junction, and a mobile dislocation with Burgers vector $\boldsymbol{b}_2$, which moves along one of the adjacent grain boundaries (Fig. 2.4d).

(III) Two grain boundary dislocations, each characterized by Burgers vector $\boldsymbol{b}$, converge into a new head dislocation with Burgers vector $2\boldsymbol{b}$ having magnitude close to the crystal lattice parameter (Fig. 2.4e). The resultant dislocation then splits into an immobile grain boundary dislocation with Burgers vector $\boldsymbol{b}_1'$, which stays at the triple junction, and a mobile lattice dislocation with Burgers vector $\boldsymbol{b}_2'$, which moves to the adjacent grain interior (Fig. 2.4f).

(IV) The head dislocation of the pile-up splits into an immobile grain boundary dislocation (with Burgers vector $\boldsymbol{b}_1$) that stays at the triple junction and a mobile partial dislocation (with Burgers vector $\boldsymbol{b}_2$) that moves in the grain interior, in which case a stacking fault is formed behind the moving partial dislocation (Fig. 2.4g).

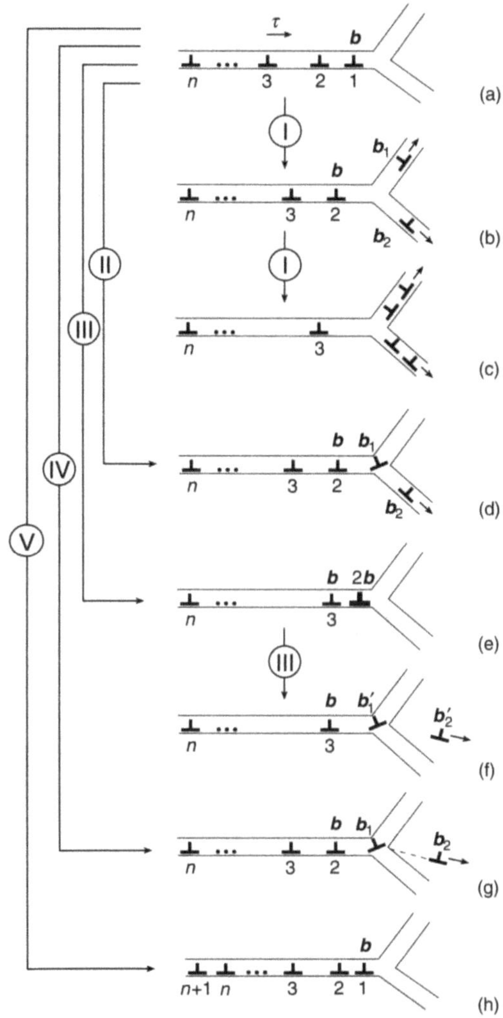

Fig. 2.4. Transformations of grain boundary dislocation pile-ups at triple junctions of grain boundaries. (**a**) Dislocation pile-up stops at triple junction. (**b**) Head dislocation of pile-up splits into two mobile dislocations that move along adjacent grain boundaries. (**c**) Second head dislocation of pile-up splits into two mobile dislocations that move along adjacent grain boundaries. (**d**) Head dislocation of pile-up splits into immobile dislocation that stays at triple junction and mobile dislocation moving along an adjacent boundary. (**e**) Two dislocations converge into a dislocation with Burgers vector $2\boldsymbol{b}$. (**f**) Dislocation with Burgers vector $2\boldsymbol{b}$ splits into immobile grain boundary dislocation that stays at triple junction and lattice dislocation that moves into grain interior. (**g**) Head dislocation of pile-up splits into immobile grain boundary dislocation that stays at triple junction and mobile partial dislocation that moves into grain interior, with stacking fault (*dashed line*) formed behind it. (**h**) New dislocation joins the stopped dislocation pile-up

(V) The pile-up is immobile. It accumulates a new grain boundary dislocation generated under the action of mechanical load (Fig. 2.4h). This process precedes one of the processes (I–IV).

The aforementioned processes of dislocation pile-up transformations were analyzed in [309] from the standpoint of their energetic preference. The latter strongly depends on the values of the angles α_1 and α_2 between the Burgers vectors $\boldsymbol{b}$ and $\boldsymbol{b}_1$, and $\boldsymbol{b}$ and $\boldsymbol{b}_2$, respectively. For example, the change in the total energy of the system ΔE_{I}, i.e., the difference between the total energies of the system in the final 'split' (Fig. 2.4b) and initial 'unsplit' (Fig. 2.4a) states, is negative for process I for $\alpha_1 = \alpha_2 = 20°$. This means the process is energetically favourable because it leads to a reduction in the total energy (see Fig. 2.5a). But it is initially positive for $\alpha_1 = \alpha_2 = 50°$. This means the process is initially energetically unfavourable due to an energetic barrier E_{mI} (see Fig. 2.5b). This difference in energy characteristics allows one to distinguish the so-called 'soft' and 'hard' triple junctions as those where the dislocation pile-up splitting does or does not occur, respectively, at given conditions of loading (say, a value of the applied stress τ). It was shown in [309] that the transition from the barrier to the barrierless regime does not in practice depend on the number n of dislocations in the pile-up and occurs at a critical angle $\alpha_{\mathrm{c}}(= \alpha_1 = \alpha_2) \approx 48°$. It is natural that both the barrier magnitude E_{mI} and corresponding critical split distance w_{cI} (Fig. 2.5b) diminish with increasing n and τ.

The other three processes of transformation of the initial dislocation pile-up (Fig. 2.4) were analysed in a similar way [309]. It turns out that the conditions favourable for process II are very close to those for process I, although the latter is still preferred. In contrast, neither process III and IV is energetically favourable for real values of their parameters. In the framework of the model [309], the ratio $F_{\mathrm{s}}/F_{\mathrm{h}}$ of the volume fraction F_{s} of soft triple junctions to the volume fraction F_{h} of hard junctions effectively characterizes the contribution of grain boundary sliding to plastic deformation. It is therefore an important structure-sensitive parameter for the behavior of NCMs under mechanical loading. Materials with a high ratio $F_{\mathrm{s}}/F_{\mathrm{h}}$ are expected to exhibit enhanced grain boundary sliding characterized by the yield stress value being lower than the yield stress values specifying alternative deformation mechanisms, e.g., diffusional creep mechanisms. Plastic flow in materials with a low ratio $F_{\mathrm{s}}/F_{\mathrm{h}}$ should occur by alternative deformation mechanisms, because grain boundary sliding is suppressed. We will discuss this question in more detail in Sect. 2.2.4.

A number of works on computer simulation of plastic deformation in model nanocrystalline f.c.c. metals have appeared over the past few years [310–315]. The competition between lattice dislocation motion and grain boundary sliding has been revealed. For small grain size ($\lesssim$ 10 nm), all deformation is accommodated in grain boundaries. At higher grain sizes, intragrain deformation is observed. This occurs via the motion of partial dislocations in grain in-

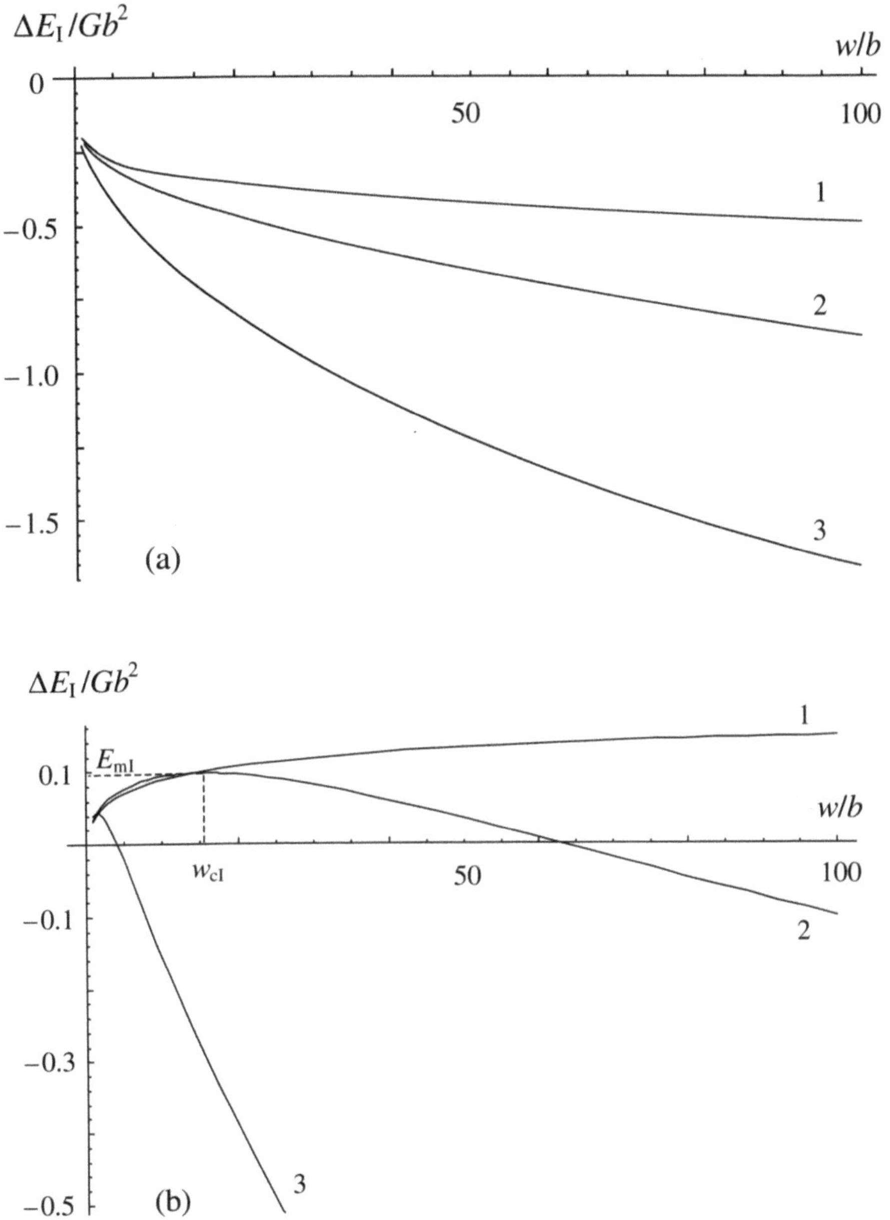

Fig. 2.5. Energy difference ΔE_{I} via the path w of gliding dislocations in the case of $n = 5$ dislocations in the initial pile-up under the external shear stresses $\tau/G = 0.0002$, 0.002, and 0.01 (*curves* 1, 2, and 3, respectively), for the Burgers vector orientations $\alpha_1 = \alpha_2 = \alpha = 20°$ (**a**) and $50°$ (**b**)

teriors, in which case these dislocations are emitted and absorbed in the opposite grain boundaries. Computer modeling results thus support the idea of competition between conventional intragrain sliding via lattice dislocation motion and grain boundary sliding in NCMs. The role of grain boundary sliding increases with falling grain size.

Song et al. [226] have distinguished two branches in the grain size dependence of the yield stress. One branch is realized at large grain sizes ($d > d_{cr}$) and described by the classical Hall–Petch relationship (2.1). At small grain sizes ($d < d_{cr}$), following the model [226], an NCM behaves as a coherent-precipitate-strengthened two-phase alloy in which the grain boundary phase plays the role of matrix, and each grain embeds coherently into the matrix. In this region, the yield stress is given as

$$\tau = \tilde{\tau} + \frac{\gamma_0}{\tilde{b}} f_{cr} , \qquad (2.32)$$

where $\tilde{\tau}$ denotes the yield stress of the matrix (grain boundary phase) without obstacles (grains), γ_0 is the interphase boundary energy, $\tilde{b}$ is the magnitude of the Burgers vector that characterizes grain boundary dislocations as carriers of plastic flow in the matrix, and f_{cr} is the volume fraction of crystalline grains (playing the role of obstacles for grain boundary dislocations), which can be written approximately as

$$f_{cr} = \left(\frac{d - \Delta}{d} \right)^3 . \qquad (2.33)$$

Here Δ is the effective grain boundary thickness, and d is the mean grain size. Equations (2.32) and (2.33) account for the experimental data on the abnormal Hall–Petch relationship at some values of the model parameters. However, use of (2.32), which initially describes the strengthening of coherent-precipitate two-phase alloys, is questionable in the case of NCMs, because it is based on the assumptions that the carriers of plastic flow in the matrix (grain boundaries) and those in the coherently embedded second-phase particles (grains) are the same. This assumption is debatable. Crystalline nanograins and grain boundary structures are essentially different: they contain dislocations with essentially different characteristics and cannot be matched coherently.

Let us also consider the model of Chokshi et al. [242]. Clearly, at sufficiently small grain sizes, the Hall–Petch model based upon the lattice dislocations may not be operative. In this region, the authors [242] have proposed room temperature Coble creep as a mechanism to explain their experimental results. Certainly, there is an order of magnitude agreement and the trend is correct. However, the functional dependence of τ on d is incorrect, as pointed out by Neih and Wadsworth [278]. Conventional Coble creep demands that $\tau \sim d^3 \propto (d^{-1/2})^{-6}$ (i.e., the τ vs. $d^{-1/2}$ curve falls very steeply as $d^{-1/2}$ increases). This is not found experimentally [242]. When we fit the data of Chokshi et al. [242] (Fig. 2.3a) to a Coble type equation, viz.,

$$\tau = \gamma + \alpha d^3 , \tag{2.34}$$

we find that the fit requires a large value of $\gamma = 360$ MPa. This suggests that the threshold stress for Coble creep is of the form Gb/d [316], since this would have the correct magnitude for a 10 nm grain size. The origin of this threshold is related to vacancies that are created and destroyed on the dislocations climbing along grain boundaries. The dislocations are pinned at grain boundary nodes and require a stress of Gb/d to climb. However, Chokshi et al. [242] showed that their data fit the relation

$$\tau = \beta - K' d^{-1/2} \tag{2.35}$$

better, with $\beta = 937$ MPa and $K' = 0.027$ MPa m$^{1/2}$ instead of (2.34). Equation (2.35) cannot be simply related to any known mechanism.

Even if the Coble creep arguments were valid for grain sizes $d < 30$ nm, we would still have to explain the behavior in the 20 to 200 nm range. This is evidently the transition regime between the Hall–Petch and Coble-creep-type behavior. The transition regime is effectively described by introducing a grain size distribution. This is the subject of the next section.

2.2.3 Competition Between Deformation Mechanisms and Effect of a Grain Size Distribution

From the previous discussion, it is quite obvious that a unified model of grain size strengthening applicable to a wide range of grain size was lacking. Masumura et al. [231] provided such a model and developed an analytical expression for τ as a function of the inverse square root of d in a simple and approximate manner that could be compared with experimental data over a wide range of grain sizes. This model is summarized below. The assumptions in this model are:

- It is assumed that polycrystals with a relatively large average grain size obey the classical Hall–Petch relation.
- At the other extreme, for very small grain sizes, it is assumed that Coble creep is active and that the τ vs. d relationship is given by

  $$\tau_c = \frac{A}{d} + Bd^3 , \tag{2.36}$$

  where B is both temperature and strain-rate dependent. The additional term A/d (the threshold term) can be large if d is in the nanometer range. For intermediate grain sizes, both mechanisms may be active if the specimen has a certain spread in its grain size distribution.
- The statistical nature of the grain sizes in a polycrystal is taken into consideration by using an analysis similar to Kurzydlowski [317]. The volume of the grains is assumed to be log-normally distributed,

$$f(v) = \frac{1}{v\sqrt{2\pi s_{\ln v}^2}} \exp\left[\frac{-(\ln v - m_{\ln v})^2}{2s_{\ln v}^2}\right] , \tag{2.37}$$

where $m_{\ln v}$ and $s_{\ln v}$ are the mean value and standard deviation of $\ln v$, respectively, and $m_{\rm v}$ is the mean volume of all the grains,

$$m_{\rm v} = \int_0^\infty v f(v) {\rm d}v = \exp\left[m_{\ln v} + \frac{(s_{\ln v})^2}{2}\right] . \tag{2.38}$$

This can also be written as $m_{\rm v} = \kappa \bar{d}^3$, where $\bar{d}$ is the mean grain size and κ is a geometrical shape factor considered for this analysis to be equal to 1.

- Finally, it is assumed that a grain size d^* exists at which the classical Hall–Petch mechanism switches to the Coble creep mechanism, $\tau_{\rm hp} = \tau_{\rm c}$ at $d = d^*$. Using (2.1) and (2.36), we have

$$K(d^*)^{-1/2} = A/d^* + B(d^*)^3 , \tag{2.39}$$

from which d^* can be determined.

Then the yield stress after averaging is given by [231]

$$\langle \tau - \tau_0 \rangle = F_{\rm hp} + F_{\rm c} , \tag{2.40}$$

where

$$F_{\rm hp} = \frac{1}{m_{\rm v}} \int_{v^*}^{\infty} \tau_{\rm hp} v f(v) {\rm d}v , \tag{2.41}$$

$$F_{\rm c} = \frac{1}{m_{\rm v}} \int_0^{v^*} \tau_{\rm c} v f(v) {\rm d}v , \tag{2.42}$$

and $v^* = (d^*)^3$.

The following definitions are used:

$$\xi = \frac{\bar{d}}{d^*} \quad \text{and} \quad \sigma = s_{\ln v} , \tag{2.43}$$

along with

$$\Lambda(\xi; \sigma, \alpha) = \xi^{3\alpha} \exp\left[\frac{\sigma^2}{2}\alpha(\alpha + 1)\right] \tag{2.44}$$

and

$$\Phi(v; \sigma, \alpha) = {\rm erf}\left\{\left(\frac{1}{2\sigma^2}\right)^{1/2}\left[\ln\left(\frac{v}{m_{\rm v}}\right) - \frac{\sigma^2}{2}(2\alpha + 1)\right]\right\} , \tag{2.45}$$

where erf{...} is the error function. Defining additional normalized variables,

$$F_{\rm hp} = K(4d^*)^{-1/2} f_{\rm hp}\,, \quad f_{\rm hp} = \Lambda(\xi;\sigma,-1/6)[1-\Phi(\xi;\sigma,-1/6)]\,, \tag{2.46}$$

$$F_{\rm c1} = A(2d^*)^{-1} f_{\rm c1}\,, \quad f_{\rm c1} = \Lambda(\xi;\sigma,-1/3)[\Phi(\xi;\sigma,-1/3)+1]\,, \tag{2.47}$$

$$F_{\rm c2} = B\frac{(d^*)^3}{2} f_{\rm c2}\,, \quad f_{\rm c2} = \Lambda(\xi;\sigma,1)[\Phi(\xi;\sigma,1)+1]\,, \tag{2.48}$$

and $P = (A/d^*)/[B(d^*)^3]$, we have a normalized form of the yield stress $\tau_{\rm n}$ as a function of the scaled grain size ξ, grain size parameter σ, and ratio P:

$$\tau_{\rm n} = \frac{2}{K(d^*)^{-1/2}}\langle \tau - \tau_0 \rangle = f_{\rm hp} + \frac{P f_{\rm c1} + f_{\rm c2}}{1+P}\,. \tag{2.49}$$

The parameter P is the ratio of Coble threshold stress to conventional stress evaluated at d^* where the transition from Coble creep to Hall–Petch strengthening occurs. For each P and σ, a universal curve is obtained, with the form and shape of the curve similar to experimental data. Equation (2.49) is the key result of the analysis given in [231]. In Figure 2.6, the yield stress for the NiP data of McMahon and Erb [318] after normalization is compared with this equation for $d^* = 5.5$ nm, $P = 0.01$, and $\sigma = 1$. The value for d^* determined from this analysis agrees with the original hardness data, where the hardness (or stress) begins to decrease with decreasing grain size at a grain size of 5–6 nm. The threshold stress as compared to the Coble creep stress is small in this material, which gives such a small value for the ratio P.

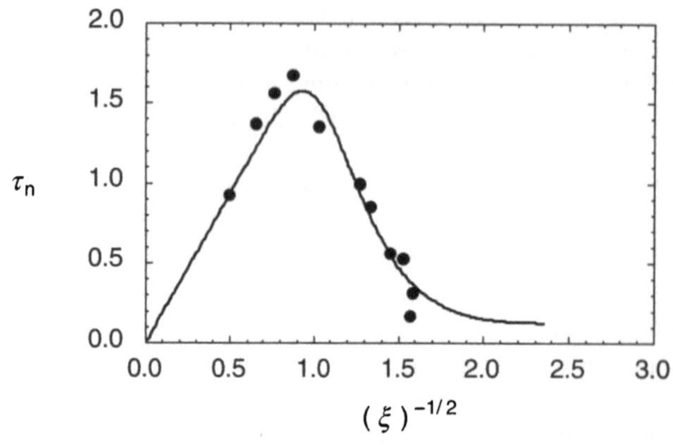

Fig. 2.6. Plot of normalized stress $\tau_{\rm n}$ as a function of the reduced variable $\xi^{-1/2}$ [231]. The experimental points for NiP alloy are taken from [318]. Values of model parameters are $d^* = 5.5$ nm, $\sigma = 1$, $p = 0.01$

Hence, the model of Masumura et al. [231] uses conventional Hall–Petch strengthening for larger grains and Coble creep with a threshold stress for smaller grains. In a material with a distribution of grain sizes, a fraction of the grains deforms by a dislocation glide process and the rest by vacancy

transport. As the average grain size decreases, the fraction deforming by glide decreases and the overall response changes from strengthening to softening. The exact form of the yield stress against grain size curve depends on the relative values of the Hall–Petch slope K, the conventional Coble constant B, the threshold constant A, and the width of the grain size distribution.

It should be noted that the model [231] only takes into account the first term of (2.31) in the calculation. However, additional terms can easily be taken into consideration and the final result will include two more terms involving error functions. Preliminary estimates made by the authors [231] indicate that this will not substantially change their result, but will make the transition from rising τ to decreasing τ (Fig. 2.6) more gradual.

Thus, following the model of Masumura et al. [231], both dislocations and Coble creep play a significant role, leading to a fuller understanding of the Hall–Petch-type relation over the whole range of grain sizes. Further, it is unrealistic to expect unlimited strength to become available simply by reducing the grain size, although in some materials, considerable strengthening can be achieved by nanocrystalline processing. Further strengthening is in principle possible, if Coble creep could somehow be reduced.

2.2.4 Grain Boundary Sliding and Triple Junction Migration

In Sect. 2.2.2, we discussed the role of grain boundary triple junctions in development of grain boundary sliding, which becomes the dominant mechanism of plasticity in NCMs under special conditions. Let us now consider how correlated sliding might occur on adjacent grain boundaries, and how the Hall–Petch-type relation might change in this case. An appropriate theoretical model has been proposed recently in [319].

Let us take a two-dimensional model of a nanocrystalline specimen fabricated in highly non-equilibrium conditions. Most fabrication routes produce grain boundaries with a high level of defects. In particular, grain boundaries with an excess density of grain boundary dislocations – carriers of grain boundary sliding – often exist in as-fabricated NCMs [28, 320] (Fig. 2.7a). When a mechanical load is applied to the specimen, mobile grain boundary dislocations (with Burgers vector parallel to grain boundary planes) are caused to move, leading to grain boundary sliding. Some moving grain boundary dislocations annihilate with others that have Burgers vectors of opposite sign (Fig. 2.7b). Others are stopped at triple junctions of grain boundaries, which represent effective obstacles for dislocation movement (Fig. 2.7b). In general, grain boundary dislocations stopped near a triple junction are capable of overcoming the junction obstacle by a dislocation reaction when the shear stress reaches some critical value. In NCMs with their high-density ensembles of triple junctions (see Sect. 1.2), the critical shear stress needed for grain boundary dislocations to overcome triple junctions specifies the contribution of grain boundary sliding to the yield stress.

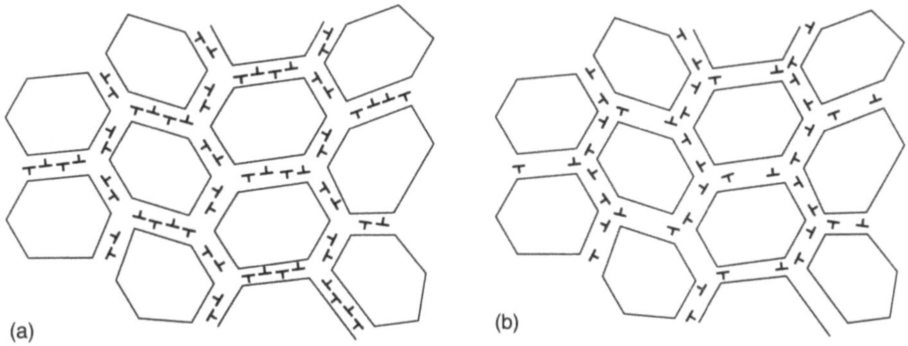

Fig. 2.7. Grain boundary dislocation ensemble in an as-prepared nanocrystalline specimen (**a**) before and (**b**) after mechanical load

Consider a model configuration of grain boundary dislocations in an NCM, formed near a triple junction under the action of shear stress τ (Fig. 2.8a) [319]. The configuration consists of three dipoles of grain boundary dislocations with Burgers vectors $\pm\boldsymbol{b}_0$, $\pm\boldsymbol{b}_1$ and $\pm\boldsymbol{b}_2$ parallel to the corresponding grain boundary planes adjacent to the triple junction. The dislocations are stopped by the (central) triple junction and its neighbouring triple junctions (Fig. 2.8a). In this case, the interspacings between the dislocations which form the three dipole configurations are approximately equal to the corresponding grain boundary lengths d_0, d_1 and d_2, respectively. Let us designate the grain boundary dislocation dipoles as A, B and C, respectively. In these considerations, we assume that energetic barriers for grain boundary dislocations to overcome triple junctions are all different, but that the lowest barrier W_{b} characterizes the central triple junction. With this assumption, in order to estimate the yield stress for grain boundary sliding, one can analyse the energetic chracteristics of transformations of grain boundary dislocations, occurring at the central triple junction. In doing so, we note that an elementary act of grain boundary sliding is a transfer of the grain boundary dislocation with Burgers vector $\boldsymbol{b}_0$ (initially stopped by the central triple junction at the grain boundary of length d_0) to the neighbouring grain boundary of length d_2. This transfer over a short distance l occurs at a critical shear stress τ_{c} at which the transfer becomes energetically favourable. The transfer is accompanied by a dislocation reaction which involves all three grain boundary dislocations stopped at the central triple junction and results in the formation of a sessile grain boundary dislocation with Burgers vector $\boldsymbol{b}_3 = \boldsymbol{b}_0 - \boldsymbol{b}_1 - \boldsymbol{b}_2$ (Fig. 2.8b). In a special case, total annihilation of all three dislocations is possible, when $b_3 = 0$. The triple junction in its turn migrates over a distance equal to b_0 under the action of the effective surface tension of the adjacent grain boundaries. Of course, within such a simplified model, no complicated transformations of atomic structure inside the triple junction which produce its migration are considered. The correspond-

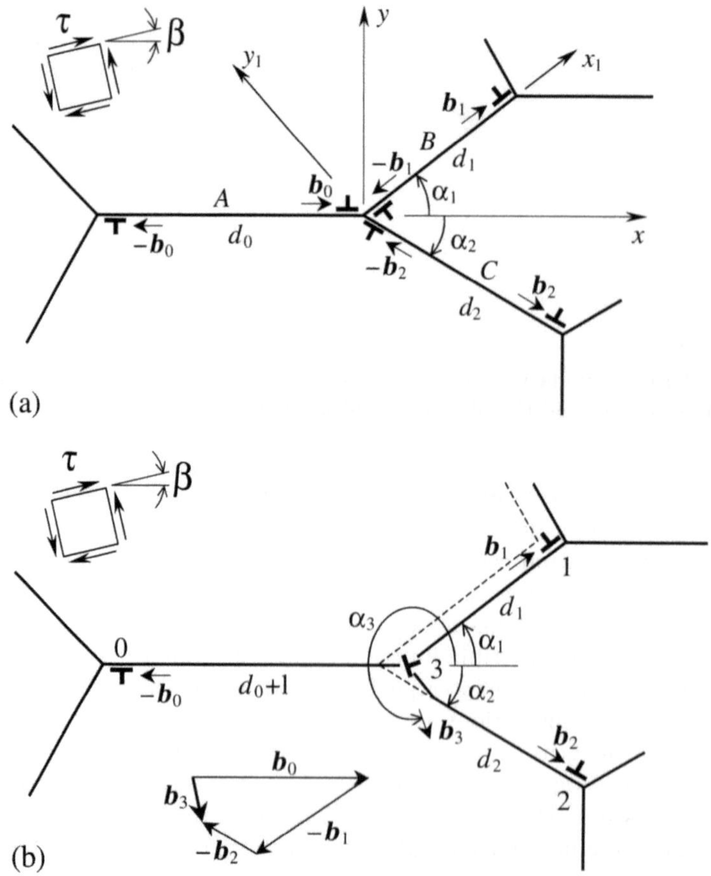

Fig. 2.8. Elementary act of grain boundary sliding in a mechanically loaded nanocrystalline specimen. (**a**) Grain boundary dislocations accumulate near triple junctions. (**b**) Dislocation with Burgers vector $\boldsymbol{b}_0$ transfers through distance l, whence it comes into reaction with two dislocations with Burgers vectors $-\boldsymbol{b}_1$ and $-\boldsymbol{b}_2$, resulting in the formation of a grain boundary dislocation with Burgers vector $\boldsymbol{b}_3$

ing energy dissipation is simply included in the magnitude of the energy barrier W_{b}.

As shown in [319], the change in the total energy (per unit length of dislocations) of the system $\Delta W^{\mathrm{t}} = W_2^{\mathrm{t}} - W_1^{\mathrm{t}}$ during its transition from the 'initial' state characterized by the energy W_1^{t} (Fig. 2.8a) into the 'final' state characterized by the energy W_2^{t} (Fig. 2.8b) may be written as

$$\Delta W^{\mathrm{t}} = W_{\mathrm{b}} - \tau b_0 l \cos(2\beta) + D\left\{ \frac{Z}{2}\left(b_3^2 - b_0^2 - b_1^2 - b_2^2\right) + b_0^2 \ln \frac{d_0 + l}{d_0} \right.$$
$$+ b_0 \sum_{i=1}^{2} \left[\frac{b_i^2}{2} \cos\alpha_i \ln \frac{d_0^2\left[(d_0+l)^2 + 2(d_0+l)d_i\cos\alpha_i + d_i^2\right]}{(d_0+l)^2(d_0^2 + 2d_0 d_i \cos\alpha_i + d_i^2)} \right.$$
$$+ \frac{b_i d_0 d_i \sin^2\alpha_i}{d_0^2 + 2d_0 d_i \cos\alpha_i + d_i^2}$$
$$\left.\left. - \frac{b_i (d_0+l) d_i \sin^2\alpha_i}{(d_0+l)^2 + 2(d_0+l)d_i\cos\alpha_i + d_i^2} \right] \right\}, \tag{2.50}$$

where $D = G/[2\pi(1-\nu)]$, G is the shear modulus, ν is the Poisson ratio, and Z is the factor taking into account the contribution of the dislocation core to the total energy. The structure of formula (2.50) for ΔW^{t} is evident. The right-hand side consists of the energy barrier (the first term) from which the work spent to move the dislocation a distance l under the shear stress τ (the second term) is subtracted, and to which the terms in curly brackets are added. The first term of the latter describes a change in the sum dislocation core energy due to the dislocation reaction at the triple junction. The second term in curly brackets specifies a change in the self-strain energy of the dipole A due to transfer of one dislocation composing the dipole through a distance l. The terms in the sum describe changes in the interaction energies associated with the dislocation transfer in question.

The transformation of the grain boundary dislocation configuration (see Fig. 2.8) occurs as an energetically favourable process, if $\Delta W^{\mathrm{t}} < 0$. In this situation, the equation $\Delta W^{\mathrm{t}} = 0$ allows one to obtain the critical shear stress τ_{c} needed to realize the transformation. In order to find an analytic relationship between τ_{c}, the dislocation Burgers vectors and the geometric parameters of triple junctions, let us first consider the special case where $d_0 = d_1 = d_2 = d$. In these circumstances, (2.50) simplifies and the critical shear stress τ_{c} is found from [319]

$$\tau_{\mathrm{c}}'(d) = p + q\frac{l}{d} + s\frac{l^2}{d^2}, \tag{2.51}$$

where

$$\tau_{\mathrm{c}}' = \frac{\tau_{\mathrm{c}}}{D} b_0 l \cos(2\beta), \tag{2.52}$$

$$p = \frac{W_{\mathrm{b}}}{D} + \frac{Z}{2}(b_3^2 - b_0^2 - b_1^2 - b_2^2), \tag{2.53}$$

$$q = \frac{b_0}{2}(2b_0 - b_1\cos\alpha_1 - b_2\cos\alpha_2), \tag{2.54}$$

$$s = \frac{b_0}{4}\left[\frac{b_1 \sin^2\alpha_1}{(1+\cos\alpha_1)^2} + \frac{b_2\sin^2\alpha_2}{(1+\cos\alpha_2)^2}\right]. \tag{2.55}$$

In deriving (2.51), (2.54) and (2.55), the inequality $l \ll d$ and the relationships $\ln(1+l/d) \approx l/d$ and $(1+l/d)^2 \approx 1+2l/d$ have been used. From Fig. 2.8b, the following relationships can be extracted:

$$b_0 = b_1 \cos\alpha_1 + b_2 \cos\alpha_2 + b_3 \cos\alpha_3 \quad \text{and} \quad b_1 \sin\alpha_1 = b_2 \sin\alpha_2 - b_3 \sin\alpha_3 \ .$$

With these relationships, the second term in (2.53) can be rewritten as

$$b_3^2 - b_0^2 - b_1^2 - b_2^2 = 2b_1 b_2 \cos(\alpha_1 + \alpha_2) - 2b_0\,(b_1 \cos\alpha_1 + b_2 \cos\alpha_2) \ . \quad (2.56)$$

With this formula, all the parameters appearing in (2.51) depend only on the characteristics of the initial defect configuration shown in Fig. 2.8a. The parameter p is always positive, but it may change in a non-monotonic way with variations in b_i and α_i (see below). The parameter q is either positive or negative, depending on the relationships between b_0 and $b_i \cos\alpha_i$ $(i = 1, 2)$. In particular, $q < 0$ for low α_i and large b_i, and $q > 0$ in the opposite case. Values of q (and, as a corollary, values of τ_c') increase with rising angles α_i and decreasing b_i. The parameter s is always positive, but the contribution of the term sl^2/d^2 to τ_c [see (2.51)] is always negligibly small.

Thus, in the situation where all grains have the same size d, the dependence of the shear stress τ_c (characterizing grain boundary sliding) on the grain size d is given by (2.51) with the two, hyperbolic and d^{-2}-type, terms sensitive to d. For positive (negative) values of the parameter q, the $1/d$ term gives rise to an increase (decrease, respectively) in τ_c with falling grain size d. The d^{-2}-type term, always with positive coefficient s, causes an increase in τ_c with reduction in grain size d. However, this term is commonly negligibly small (due to the low l) compared with the hyperbolic term.

With $x = d^{-1/2}$, using (2.51), the authors [319] have obtained the following dependence, written in variables used to describe the classical Hall–Petch dependence:

$$\tau_c'(x) = p + qlx^2 + sl^2x^4 \ . \quad (2.57)$$

This formula is convenient for analyzing deviations of the $\tau_c(d^{-1/2}) = \tau_c(x)$ dependence from the classical Hall–Petch relationship. From (2.57), it follows that there are two, parabolic and x^4-type terms sensitive to $x = d^{-1/2}$. The x^4-term is low, and will therefore be hard to detect experimentally, in contrast to the parabolic term. The x^2 term, in the framework of our model, is responsible for deviations of the $\tau_c(d^{-1/2})$ dependence from the classical Hall–Petch relationship.

Now let us consider the general situation where grain boundaries adjacent to the central triple junction have different length values. To describe this situation, we suppose $d_i/d_0 = a_i$ $(i = 1, 2)$ and $d_0 = d$. In these circumstances, the first term p in (2.57) does not change, while the parameter q appearing in the second term on the right-hand side of (2.57) has the form [319]

$$q(a_1, a_2) = b_0 \left[b_0 - b_1 \frac{a_1 \cos\alpha_1 (a_1 + \cos\alpha_1)}{a_1^2 + 2a_1 \cos\alpha_1 + 1} - b_2 \frac{a_2 \cos\alpha_2 (a_2 + \cos\alpha_2)}{a_2^2 + 2a_2 \cos\alpha_2 + 1} \right] . \tag{2.58}$$

The sign of the parameter q given by (2.58) is highly sensitive to the parameters a_i.

The third term on the right-hand side of (2.57) is [319]

$$S(x, a_1, a_2) = b_0 \sum_{i=1}^{2} \left[\frac{b_i a_i \sin^2 \alpha_i}{a_i^2 + 2a_i \cos\alpha_i + 1} - \frac{b_i a_i \sin^2 \alpha_i (1 + lx^2)}{a_i^2 + 2a_i \cos\alpha_i (1 + lx^2) + (1 + lx^2)^2} \right] . \tag{2.59}$$

Thus, in the situation discussed, the critical shear stress that characterizes grain boundary sliding in the framework of this model is represented as

$$\tau_c'(x, a_1, a_2) = p + q(a_1, a_2) lx^2 + S(x, a_1, a_2) , \tag{2.60}$$

where p, $q(a_1, a_2)$ and $S(x, a_1, a_2)$ are given by (2.53), (2.58) and (2.59), respectively. This is the main result of the model [319].

The authors [319] compared their results with the experimental data of Volpp et al. [321] on measurements of the microhardness H_v in as-prepared and heat-treated nanocrystalline NiAl materials synthesized by a ball-milling technique. The following characteristic values of parameters of these materials and the defect configuration under consideration were used in the numerical calculations. The elastic constants were taken from [321] as $G = 70$ GPa and $\nu = 0.314$. For simplicity, it was assumed that $\alpha_1 = \alpha_2 = \alpha$, $Z = 1$ and $\beta = 0$. The moduli of the grain boundary dislocation Burgers vectors were chosen as $b_0 = 0.1$ nm, $b_1 = b_2 = 0.08$ nm. The distance through which the grain boundary dislocation with Burgers vector $\boldsymbol{b}_0$ moves was approximated by $l =$ 1 nm, which corresponds to the characteristic thickness of grain boundaries. The energetic barrier for grain boundary dislocation motion was assumed to be $W_b = \xi G b_0^2$ (per unit length of dislocation), where ξ is a fitting parameter of the order of unity. In order to compare theoretical calculations of the yield stress τ_c with experimental measurements of the microhardness H_v, the authors [319] used the known relationship [269]: $H_v^{th} \approx 3\sigma_y(\varepsilon = 8\%) \approx 6\tau_y$.

Equation (2.60) predicts the relation between yield stress and grain size provided that the mechanism described here is dominant. To compare this result with experiments, especially with those of Volpp et al. [321], one can proceed by noting that the last term, i.e., $S(x, a_1, a_2)$ can be safely neglected, as mentioned before. The term $q(a_1, a_2)$ can in principle be either positive or negative, as given by (2.58). It is easy to show that, for the range of $\cos\alpha$, the following simple approximation is valid [319]:

$$q(a_1, a_2) \approx b_0 \left[b_0 - b_1 \cos\alpha \frac{a_1}{a_1 + 1} - b_2 \cos\alpha \frac{a_2}{a_2 + 1} \right] . \tag{2.61}$$

It was also shown in [319] that the average value of the ratio $a\cos\alpha/(a+1)$ can be estimated as $\langle a\cos\alpha/(a+1)\rangle \approx 3/8$. Therefore, if one takes $b_0 = 0.1$ nm and $b_1 = b_2 = 0.08$ nm as previously stated, the average $\langle q(a_1, a_2)\rangle > 0$, and even the random $q(a_1, a_2) > 0$ if a_1 and a_2 are not too large. Hence, the authors [319] used the approximation $\tau_c'(x, a_1, a_2) \approx p + q(a_1, a_2)lx^2$ with positive $q(a_1, a_2)$ given by (2.61), and then applied the procedure used by Masumura et al. [231] for averaging over the grain size distribution to take into account the range of grain sizes in an NCM (see Sect. 2.2.3 for details). This was done by assuming that the mechanism discussed in the present section operates for grain sizes less than d^* and that the usual Hall–Petch mechanism operates for grain sizes higher than d^*. Figure 2.9 gives the full curve following this procedure for the whole range of average grain sizes, where d^* was taken as 16 nm [319]. The curve then needs the value of a single parameter $\lambda = (ql)/(pd^*)$. The best fit (see the dashed curve and empty squares in Fig. 2.9) is obtained for $\lambda = 0.10$ which corresponds to $q/p = 1.6$ if l is taken to be 1 nm. This ratio appears to be reasonable, but cannot yet be calculated exactly, because of the uncertainty in the values of the parameters needed to calculate p and q.

Now let us discuss the specific features of plastic deformation in annealed NCMs. First, thermal annealing changes the structure of grain boundaries and their triple junctions, which become closer to equilibrium. The values of angles between grain boundaries at the triple junctions tend to the average 120°. Therefore, the volume fraction of 'hard' triple junctions increases (see Sect. 2.2.2) and this hampers the consecutive grain boundary gliding on adjacent boundaries. Second, heat treatment causes annihilation of grain boundary dislocations, carriers of grain boundary sliding, which also hampers grain boundary sliding. In fact, generation of new grain boundary dislocations commonly requires intensive flows of lattice dislocations from grain interiors to grain boundaries, where lattice dislocations split into grain boundary dislocations. At the same time, the density of lattice dislocations is low in annealed materials, while their nucleation under mechanical load is hampered due to nanoscale effects (see Sect. 1.2). Since both dislocation mechanisms of plasticity (the glide of lattice dislocations and grain boundary sliding) are suppressed in at least the smaller grains of an NCM, alternative deformation mechanisms are capable of effectively contributing to plastic flow or even being dominant in such materials. In this case the model proposed by Masumura et al. [231] is based on the idea that, at least for very small well-annealed grains, Coble-creep-type behavior may be applicable (see Sect. 2.2.3). If one takes $d^* = 16$ nm and a value of 0.13 for P, the continuous curve $H_v(d^{-1/2})$ in Fig. 2.9 is obtained for well-annealed samples of nanocrystalline NiAl [319]. As can be seen, this curve agrees well with the experimental points taken from [321] (black squares in Fig. 2.9) and describes both the rising and falling branches of the dependence $H_v(d^{-1/2})$ with some success.

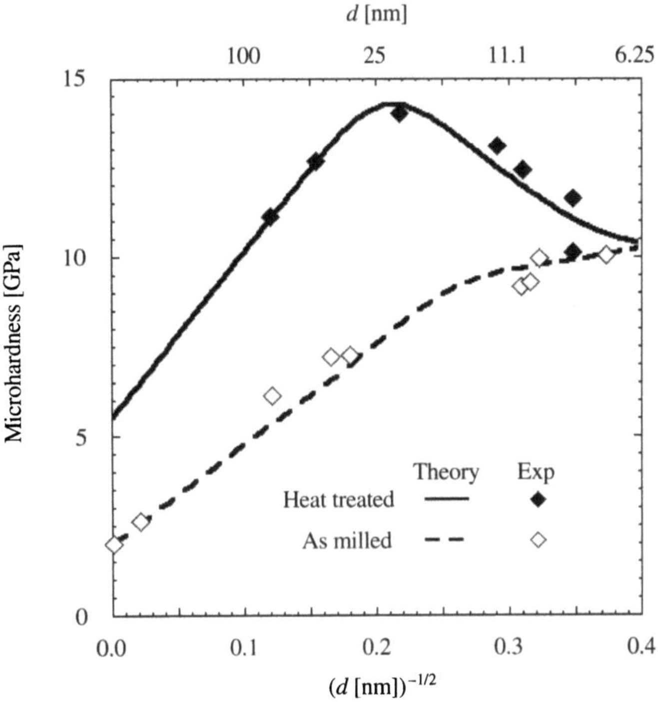

Fig. 2.9. Microhardness of nanocrystalline NiAl in as-prepared (*empty squares* from experiment [321] and *dashed curve* from theory [319]) and heat-treated (*black squares* from experiment [321] and *continuous curve* from theory [319]) states, viz., the inverse square root of the average grain size

Thus, the simple model [319] of grain boundary sliding and elementary migration of grain boundary triple junctions allows one to obtain the approximate (2.60). After some simplification and accounting for the statistical distribution of grain sizes, the equation gives a dependence of hardness on grain size that corresponds well with test results for as-prepared nanocrystalline NiAl samples [321]. It has been shown that this model cannot be used to describe the mechanical behavior of thermally annealed samples of the same NCM because here grain boundary sliding is supressed due to structural transformations of the grain boundaries and their triple junctions, combined with an abrupt decrease in density of both the lattice and grain boundary dislocations as well. In this case, suitable models are those which account for mechanisms of diffusional plasticity on grain boundaries (Coble creep) [231, 319].

2.2.5 Triple Junction Diffusion and Abnormal Hall–Petch Dependence

In the development of the approach by Masumura et al. [231], a theoretical model [211] has recently been suggested, describing a contribution of the deformation mechanism associated with triple junction diffusion to plastic flow in NCMs. As with grain boundary diffusion, triple junction diffusion is capable of playing a very important role in plastically deformed NCMs, where the volume fraction of triple junctions of grain boundaries is extremely high (see Sect. 1.2). In [211], the abnormal Hall–Petch effect in NCMs has been considered to be based upon competition between conventional lattice dislocation glide, grain boundary diffusional creep and triple junction diffusional creep. The key statements of the model [211] are as follows:

1. Materials with a relatively large grain size obey the classical Hall–Petch relationship (2.1).
2. Plastic deformation in materials with a small grain size occurs via diffusional creep mechanisms provided by enhanced diffusion along grain boundaries and their triple junctions. In this situation, the plastic strain rate $\dot{\varepsilon}$ has two basic constituents [211]:

$$\dot{\varepsilon} = \dot{\varepsilon}_{\mathrm{c}} + \dot{\varepsilon}_{\mathrm{tj}} , \tag{2.62}$$

where $\dot{\varepsilon}_{\mathrm{c}}$ is the plastic strain rate associated with grain boundary diffusional creep (Coble creep) and $\dot{\varepsilon}_{\mathrm{tj}}$ is the plastic strain rate associated with triple junction diffusional creep, considered in detail by Rabukhin [203]. The Coble creep strain rate is taken as (e.g., [322]):

$$\dot{\varepsilon}_{\mathrm{c}} = 14\pi \frac{\Omega \sigma w}{k_{\mathrm{B}} T} \frac{D_{\mathrm{gb}}}{d^3} , \tag{2.63}$$

where k_{B} denotes the Boltzmann constant, T the absolute temperature, Ω the atomic volume, w the grain boundary thickness, σ the applied normal stress, and D_{gb} the coefficient of grain boundary self-diffusion. The creep associated with enhanced triple junction diffusion is characterized by the plastic strain rate [203]:

$$\dot{\varepsilon}_{\mathrm{tj}} = 20 \frac{\Omega \sigma l \delta}{k_{\mathrm{B}} T} \frac{D_{\mathrm{tj}}}{d^4} , \tag{2.64}$$

where D_{tj} is the coefficient of triple-junction self-diffusion, δ is the triple junction tube diameter, and l is the characteristic length of distortions induced by a triple junction in its adjacent grain boundaries. Substituting (2.63) and (2.64) into (2.62), we have the following formula for the yield stress, which characterizes the diffusional deformation mechanisms in materials with small grain size [211]:

$$\sigma_{\mathrm{diff}} = \frac{\dot{\varepsilon} d^4 k_{\mathrm{B}} T}{2\Omega (10\delta l D_{\mathrm{tj}} + 7\pi w d D_{\mathrm{gb}})} . \tag{2.65}$$

3. The statistical nature of the grain size (see [231] and Sect. 2.2.3) is taken into account in a fine-grained material. Following [231], the volume v of

the grains is assumed to be log-normally distributed as given by (2.37), with the mean volume m_v of all the grains given by (2.38). The yield stress σ is thus determined, not from the average grain size, but from the whole distribution.

4. Finally, it is assumed that a grain size d^* exists at which the classical Hall–Petch mechanism switches to the diffusional mechanisms, $\sigma_{\mathrm{hp}} = \sigma_{\mathrm{diff}}$ at $d = d^*$. Using (2.1) and (2.65) with $d = d^*$, Fedorov et al. [211] obtained the equation

$$\sigma_0 + K(d^*)^{-1/2} = \frac{\dot{\varepsilon}(d^*)^4 k_{\mathrm{B}} T}{2\Omega(10\delta l D_{\mathrm{tj}} + 7\pi w d^* D_{\mathrm{gb}})} \tag{2.66}$$

for d^*. With the grain size distribution (2.37), the yield stress reads

$$\langle \sigma_{\mathrm{tot}} \rangle = F_{\mathrm{hp}} + F_{\mathrm{diff}} \ , \tag{2.67}$$

where

$$F_{\mathrm{hp}} = \frac{1}{m_{\mathrm{v}}} \int_{v^*}^{\infty} \sigma_{\mathrm{hp}} v f(v) \mathrm{d}v \ , \tag{2.68}$$

$$F_{\mathrm{diff}} = \frac{1}{m_{\mathrm{v}}} \int_{0}^{v^*} \sigma_{\mathrm{diff}} v f(v) \mathrm{d}v \ , \tag{2.69}$$

and $v^* = (d^*)^3$.

For characteristic copper parameters [203, 231, 282, 286, 322], that is, $D_{\mathrm{gb}} \approx 2.6 \times 10^{-20}\ \mathrm{m^2 s^{-1}}$, $\Omega = 1.18 \times 10^{-29}\ \mathrm{m^3}$, $k_{\mathrm{B}} = 1.38 \times 10^{-23}\ \mathrm{J\,K^{-1}}$, $T = 290\ \mathrm{K}$, $\sigma_0 = 200$ MPa, $K = 1750\ \mathrm{MPa\,nm^{1/2}}$, $s = 1$, $w = 2b$, $l = 5b$, $\delta = 3b$, and $b = 0.3$ nm (where b is the crystal lattice parameter), the dependence $\sigma(d^{-1/2})$ has been calculated numerically using (2.67)–(2.69) [211]. The value of d^* given by (2.66) turns out to be ≈ 18 nm. This estimate is close to the one made by Masumura et al. ($d^* = 15$ nm) [231]. Curves $\sigma(d^{-1/2})$ taking into account only the grain boundary diffusional creep (dashed curve) and both the grain boundary and triple junction diffusional creep mechanisms (continuous curve) are shown in Fig. 2.10 [211]. For $d \geq 20$ nm, the dashed and continuous curves are close. At the same time, triple junction diffusional creep causes an essential contribution to the high deviations of the $\sigma(d^{-1/2})$ dependence from the classical Hall–Petch relationship (2.1) (dotted line in Fig. 2.10) at small values of the grain size d. Similar deviations are inherent in the experimentally observed dependence [224–226, 233, 242, 281, 286, 287] for mechanical characteristics (yield stress, microhardness) of real fine-grained copper materials (see Fig. 2.10).

To summarize our discussion of the role of different mechanisms of plastic deformation in formation of Hall–Petch relationship for NCMs, we can conclude as follows:

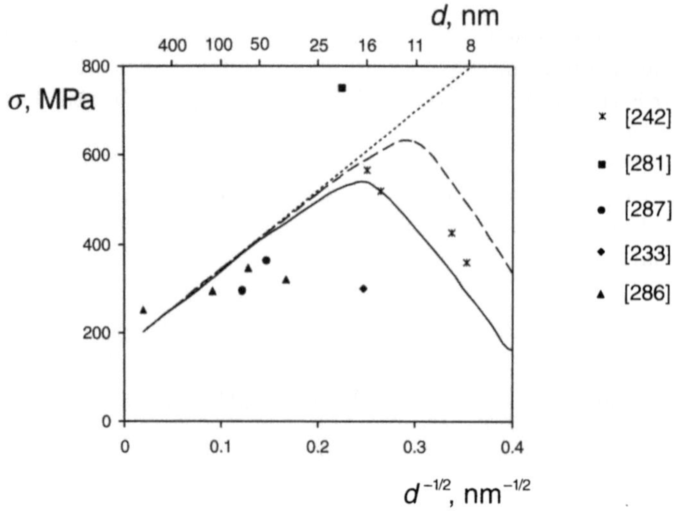

Fig. 2.10. Yield stress σ as a function of the inverse square root of the average grain size d in copper. Experimental data [233, 242, 281, 286, 287] are shown along with calculated values. *Continuous* and *dashed curves* correspond to model calculations with the contribution of triple junction diffusional creep taken into account and not taken into account, respectively. The classical Hall–Petch dependence is shown by the *dotted line*

- The competition between different physical mechanisms of plastic flow effectively accounts for the experimentally detected abnormal Hall–Petch dependence in NCMs.
- Such competing physical mechanisms of plastic flow as lattice dislocation motion in grain interiors, grain boundary sliding and diffusion plasticity mechanisms related to bulk, grain boundary and triple junction diffusion, can contribute to the processes of plastic deformation in NCMs. The competition depends strongly on structural and material characteristics as well as mechanical loading parameters.
- Each of the aforementioned deformation mechanisms can be dominant in a certain range of parameters characterizing a mechanically loaded NCM. A transition from one dominant deformation mechanism to another with reduction in the grain size gives rise to changes in the exhibited deformation behavior, in particular, deviations from the Hall–Petch dependence.
- To understand the details of the competition between the different physical mechanisms of plastic flow in NCMs, further theoretical and experimental work is needed in this area.

3 Localization of Plastic Flow in Nanocrystalline Materials

In general, plastic deformation in NCMs can be spatially homogeneous or localized in narrow shear bands. Evidence of plastic flow localization has been observed in NCMs tested in tension [244], compression [257], fatigue [323], and microhardness [324, 325]. For instance, during microhardness testing, nanostructured TiB_2 films have exhibited plastic flow which has been realized in either the homogeneous or inhomogeneous regime [324, 325]. In the latter case, their plastic behaviour has been very similar to the well-known phenomena of superplasticity which is quite common for fine-grained metals, intermetallics and ceramics under certain specific temperature and strain rate conditions [326–332]. The effect of superplasticity in NCMs has also been detected experimentally [227, 228, 333–338]. It has been well established [326–332] that grain boundary sliding is mainly responsible for the superplasticity.

During the past few years, the concept of cooperative grain boundary sliding (CGBS) has been proposed [339, 340] and supported by many further experiments [331, 332, 341–346]. In the CGBS process, groups of grains move as a unit along some gliding surfaces which contain many grain boundaries. Such cooperative shear has two main features. Firstly, it is sequential and nonuniform along the shear surfaces [342, 343] and, secondly, its progress is allowed by different accommodation mechanisms [341–343] including intragranular deformation, grain boundary migration, grain rotation, etc.

3.1 Concept of Cellular Dislocation

To describe the sequential sliding of grains, Morral and Ashby [347] introduced a model of cellular dislocations, i.e., dislocations of the grain cellular network (lattice), which have a core size and Burgers vector value of about the grain size. By analogy with lattice dislocations in crystals, which produce the sequential shear of the crystalline lattice along glide planes, the gliding of cellular dislocations produces the sequential shear of grain groups along their glide surfaces.

Zelin and Mukherjee [341], and more recently Zelin et al. [348] applied this idea to the case of CGBS. However, the problem is how to use the standard models of dislocations [246, 349] in this approach since it is well known that these models lead to elastic fields which are singular at the dislocation line. This means, for example, that the elastic strains and stresses will achieve abnormally high levels within the core of such a cellular dislocation [341,350].

In [351], cellular dislocations have been considered with the help of non-singular solutions for dislocations within the theory of gradient elasticity [149–151, 153, 154, 158], whereby one can avoid the standard singularity problem of the classical theory (see also Sect. 1.1). In this approach, the model [351] treats the homogeneous regime of plastic flow as resulting from noncorrelated gliding of the cellular dislocations and the inhomogeneous regime as resulting from correlated gliding along neighbouring grain boundary surfaces. Let us consider the application of this model to nanostructured ceramic films under microhardness testing.

When an indenter acts on the surface of a nanostructured ceramic film, in the initial instants, an elastically strained state is formed which is well known for a spherical or conical indenter [352]. During the subsequent loading, various mechanisms of elastic strain relaxation may operate depending on both the testing conditions and the material structure [353]. Andrievskii et al. [324, 325] have reported that, in the case of nanostructured Ti–B–N films, the generation of shear bands gives the main contribution to such a relaxation. On the basis of these observations, the gliding loops of cellular dislocations are assumed to be generated near the indenter tip and to extend along the planes of maximal shear stresses [351]. In other words, big groups of nanograins glide as units along grain boundary surfaces, thus producing a sequential plastic shear. The regions of the plastic shears are bounded by the lines of cellular dislocation loops which glide at the heads of such nanograin groups. One can now describe several variants of the system evolution operating with these cellular dislocation loops. Before doing so, let us consider the model for a cellular dislocation loop in more detail.

3.1.1 Cellular Dislocations in Gradient Elasticity

Let a cellular dislocation loop be characterized by its Burgers vector $\boldsymbol{B}$ and average size L, which satisfy the following relations: $|\boldsymbol{B}| = B \approx d$ and $L \gg d$, where d is an average nanograin size. We can then consider every loop segment to be a straight cellular dislocation of mixed type and subdivide it into edge and screw cellular dislocations. The main point here is an application of the gradient elasticity approach to model the elastic fields of such cellular dislocations. Consider first the governing equations of the gradient elasticity.

3.1.1.1 Governing Equations of Gradient Elasticity

In 1965, Mindlin [354] proposed a linear theory describing deformation of elastic solids in which the density of strain energy was a function of the strain and its first and second gradients. Taking into account the second gradient of the strain, the author claimed to incorporate both cohesive forces and surface tension into the linear elasticity. The corresponding modification of Hooke's law reads [354]

$$\boldsymbol{\sigma} = \lambda \left(\boldsymbol{I} - c_1 \boldsymbol{I} \nabla^2 - c_2 \nabla\nabla \right) (\operatorname{tr} \boldsymbol{\varepsilon}) + 2\mu \left(1 - c_3 \nabla^2 \right) \boldsymbol{\varepsilon} \,, \tag{3.1}$$

where λ and μ are the usual Lamé constants, $\boldsymbol{\sigma}$ and $\boldsymbol{\varepsilon}$ denote the elastic stress and strain tensors, $\boldsymbol{I}$ is the unit tensor, ∇^2 is the Laplacian, and c_1, c_2 and c_3 are three independent gradient coefficients.

Substituting (3.1) into the usual equilibrium equation $\nabla \cdot \boldsymbol{\sigma} = 0$ gives the following equation for the displacement vector $\boldsymbol{u}$ [152]:

$$\mu \left(1 - c_3 \nabla^2 \right) \nabla^2 \boldsymbol{u} + \left\{ \lambda + \mu - \left[\lambda (c_1 + c_2) + \mu c_3 \right] \nabla^2 \right\} \nabla\nabla \cdot \boldsymbol{u} = 0 \,. \tag{3.2}$$

In the case where the displacement vector $\boldsymbol{u}$ has more than one non-vanishing component, (3.2) gives a system of coupled partial differential equations which seems hard to solve. However, if the relation $c_1 + c_2 = c_3 = c$ holds, (3.2) results in

$$\left(1 - c \nabla^2 \right) \left[\mu \nabla^2 \boldsymbol{u} + (\lambda + \mu) \nabla\nabla \cdot \boldsymbol{u} \right] = 0 \,. \tag{3.3}$$

In the special case when $c_1 = c_3 = c$ and $c_2 = 0$, (3.1) transforms into

$$\boldsymbol{\sigma} = \lambda (\operatorname{tr} \boldsymbol{\varepsilon}) \boldsymbol{I} + 2\mu \boldsymbol{\varepsilon} - c \nabla^2 \left[\lambda (\operatorname{tr} \boldsymbol{\varepsilon}) \boldsymbol{I} + 2\mu \boldsymbol{\varepsilon} \right] \,. \tag{3.4}$$

This equation was initially proposed by Altan and Aifantis [355] to eliminate the strain singularity at the mode III crack tip. They also showed that, for an atomic lattice, the gradient coefficient c can be estimated [355] by $\sqrt{c} \approx a/4$, where a is the lattice constant. Substituting (3.4) into the equilibrium equation $\nabla \cdot \boldsymbol{\sigma} = 0$ leads again to (3.3). A physical derivation of (3.3) and (3.4) based on a mixture-like model for composite materials was provided by Aifantis [356] and later in more detail by Altan and Aifantis [357]. Ru and Aifantis [358] found a simplified way to solve boundary-value problems in this special theory of gradient elasticity described by (3.3) or (3.4), by reducing them to a non-homogeneous Helmholtz equation with the 'source' term given in terms of well-known solutions for the same problems in classical elasticity. They have also shown that the stress field in this theory of gradient elasticity remains the same as in classical elasticity. Altan and Aifantis [357] used a Fourier transform procedure to solve (3.3) in two dimensions, including the mode I and II crack problems.

Application of this theory to crack problems [355–361] has resulted in the elimination of classical singularities from the solutions for the elastic displacement and strain fields at the crack tips. Although the stress components remain as in the classical theory, this difficulty has been considered less severe than the strain singularity because the stress may not be rigorously defined at the atomic level near a discontinuity.

Encouraged by these results, Gutkin and Aifantis [149, 150, 152] applied the special gradient elasticity theory given by (3.4) to dislocations [149, 150] and disclinations [152]. As was the case with cracks, new gradient solutions for displacement [149, 150], strain fields [149, 150] and energies [152] of dislocations as well as for strain fields [152] of disclinations were non-singular at the defect lines. The corresponding stress fields were the same as in the classical theory of elasticity.

A further step was made by Ru and Aifantis in unpublished work [362] (see also [363]), where a simple extension of the gradient elasticity model given by (3.4) was used to dispense with both the strain and the stress singularity at the dislocation core and at the crack tip. The constitutive equation of this theory reads

$$\left(1 - c_1\nabla^2\right)\boldsymbol{\sigma} = \left(1 - c_2\nabla^2\right)\left[\lambda(\operatorname{tr}\boldsymbol{\varepsilon})\boldsymbol{I} + 2\mu\boldsymbol{\varepsilon}\right] , \tag{3.5}$$

with two different gradient coefficients c_1 and c_2. A rather simple mathematical procedure analogous to the one contained in [358] was outlined in [362] for the solution of (3.5) in terms of the classical elasticity solutions for the same boundary-value problem. In fact, it is easily established [358] (see also [149, 150, 152]) that the right-hand side of (3.5) for the case $c_1 \equiv 0$ gives the classical solution for the stress field, which we denote here by $\boldsymbol{\sigma}^0$, while the solution for the displacement is determined through the inhomogeneous Helmholtz equation given by

$$\left(1 - c_2\nabla^2\right)\boldsymbol{u} = \boldsymbol{u}^0 , \tag{3.6}$$

where $\boldsymbol{u}^0$ denotes the classical elasticity solution for the same traction boundary-value problem. Equation (3.6) implies a similar equation for the strain $\boldsymbol{\varepsilon}$ of the gradient theory:

$$\left(1 - c_2\nabla^2\right)\boldsymbol{\varepsilon} = \boldsymbol{\varepsilon}^0 , \tag{3.7}$$

in terms of the strain $\boldsymbol{\varepsilon}^0$ of the classical elasticity theory for the same traction boundary-value problem. With the displacement or strain field thus determined (which is obviously independent of whether $c_1 \equiv 0$ or $c_1 \neq 0$), it follows that the stress field $\boldsymbol{\sigma}$ of (3.5) can be determined (for the case $c_1 \neq 0$) from the equation

$$\left(1 - c_1\nabla^2\right)\boldsymbol{\sigma} = \boldsymbol{\sigma}^0 , \tag{3.8}$$

where $\boldsymbol{\sigma}^0$ denotes the solution obtained for the same boundary-value problem within the classical theory of elasticity.

In order to solve (3.5), one can thus solve (3.7) and (3.8) separately using the classical solutions $\boldsymbol{\varepsilon}^0$ and $\boldsymbol{\sigma}^0$, provided that appropriate care is taken for the extra (due to the higher order terms) boundary conditions or conditions at infinity. For dislocations and disclinations in a homogeneous medium, solutions are accounted for by assuming that the strain and stress fields at infinity are the same for both the gradient and the classical theory. The approach was first applied [362] to the cases of screw dislocations and mode III cracks. The asymptotic solutions at the dislocation line and crack tip were found, demonstrating the elimination of both strain and stress singularities there. Recently, the gradient elasticity described by (3.5) has been used to find non-singular solutions for stress fields of dislocations [151,154] and disclinations [153, 154] in a homogeneous solid. The boundary-value problems of dislocations near interphase boundaries within the gradient elasticity theory (3.5) have been solved in [155–157] where non-singular expressions have been found for dislocation stress fields as well as for 'image' forces on dislocations due to interfaces.

It is worth noting that an equation similar to (3.8) was derived earlier by Eringen [141–143] in his version of the nonlocal theory of elasticity. Recently, Lazar [162–167] has obtained similar equations for both the stress and strain fields within the framework of his gauge theory of elastoplasticity.

3.1.1.2 A General Solution for Straight Dislocations

Consider a mixed dislocation whose line coincides with the z-axis of a Cartesian coordinate system. Let its Burgers vector be $\boldsymbol{b} = b_x \boldsymbol{e}_x + b_z \boldsymbol{e}_z$ thus determining the edge (b_x) and screw (b_z) components.

Classical Solution

In the framework of classical elasticity theory, the total displacement field is described by

$$\boldsymbol{u}^0 = \frac{b_x \boldsymbol{e}_x + b_z \boldsymbol{e}_z}{2\pi} \left\{ \arctan\frac{y}{x} + \frac{\pi}{2} \mathrm{sign}(y) \left[1 - \mathrm{sign}(x)\right] \right\} + \frac{b_x}{4\pi(1-\nu)} \left\{ \boldsymbol{e}_x \frac{xy}{r^2} - \boldsymbol{e}_y \left[(1-2\nu) \ln r + \frac{x^2}{r^2} \right] \right\} , \tag{3.9}$$

where ν is the Poisson ratio and $r^2 = x^2 + y^2$. Here we use a single-valued discontinuous form suggested by de Wit [364]. The elastic strain field ε_{ij}^0 reads, in units of $1/[4\pi(1-\nu)]$ [349, 364],

$$\varepsilon_{xx}^0 = -b_x y \frac{(1-2\nu)r^2 + 2x^2}{r^4} , \quad \varepsilon_{yy}^0 = -b_x y \frac{(1-2\nu)r^2 - 2x^2}{r^4} ,$$
$$\varepsilon_{xy}^0 = b_x x \frac{x^2 - y^2}{r^4} , \quad \varepsilon_{xz}^0 = -b_z \frac{(1-\nu)y}{r^2} , \quad \varepsilon_{yz}^0 = b_z \frac{(1-\nu)x}{r^2} , \tag{3.10}$$

and the elastic stress field σ_{ij}^0 is (in units of $D = G/[2\pi(1-\nu)]$, where $G \equiv \mu$ is the shear modulus as before) [349, 364]

$$\sigma_{xx}^0 = \varepsilon_{xx}^0(\nu = 0) , \quad \sigma_{yy}^0 = \varepsilon_{yy}^0(\nu = 0) , \quad \sigma_{zz}^0 = \nu(\sigma_{xx}^0 + \sigma_{yy}^0) ,$$
$$\sigma_{xy}^0 = \varepsilon_{xy}^0 , \quad \sigma_{xz}^0 = \varepsilon_{xz}^0 , \quad \sigma_{yz}^0 = \varepsilon_{yz}^0 . \tag{3.11}$$

Fields (3.9) (y-component), (3.10) and (3.11) are singular at the dislocation line.

The elastic energy W^0 of the dislocation per unit dislocation length is [349]

$$W^0 = \frac{D}{2}\left[b_x^2 + (1-\nu)b_z^2\right] \ln \frac{R}{r_0} , \tag{3.12}$$

where R denotes the size of the solid and r_0 is a cutoff radius for the dislocation elastic field near the dislocation line. When $r_0 \to 0$, W^0 becomes singular.

Gradient Solution

Let us now consider the corresponding dislocation fields within the theory of gradient elasticity given by (3.5). As described in Sect. 3.1.1.1, one can obtain the solution of (3.5) by solving (3.6)–(3.8) separately. They can be solved using the Fourier transform method [150–157]. Omitting the intermediate calculations, we give here only the final results. For the total displacements, the solution of (3.6) gives [149–151, 154]

$$\begin{aligned} \boldsymbol{u} = \boldsymbol{u}^0 &- \frac{b_x}{4\pi(1-\nu)}\left\{\left[\boldsymbol{e}_x 2xy + \boldsymbol{e}_y(y^2 - x^2)\right] r^2 \Phi_2 + \boldsymbol{e}_y \Phi_0\right\} \\ &+ \frac{b_x \boldsymbol{e}_x + b_z \boldsymbol{e}_z}{2\pi} \operatorname{sign}(y) \int_0^{+\infty} \frac{s \sin(sx)}{1/c_2 + s^2} \exp\left(-|y|\sqrt{\frac{1}{c_2} + s^2}\right) \mathrm{d}s , \end{aligned} \tag{3.13}$$

where $\boldsymbol{u}^0$ is given by (3.9),

$$\Phi_0 = (1-2\nu) K_0(r/\sqrt{c_2}) , \quad \Phi_2 = \frac{2c_2/r^2 - K_2(r/\sqrt{c_2})}{r^4} ,$$

with $K_n(r/\sqrt{c_2})$ the modified Bessel function of the second kind and $n = 0, 1, \ldots$ the order of this function.

For the elastic strain, solution of (3.7) gives [149–151, 154] $\varepsilon_{ij} = \varepsilon_{ij}^0 + \varepsilon_{ij}^{\mathrm{gr}}$, where ε_{ij}^0 are given by (3.10) and $\varepsilon_{ij}^{\mathrm{gr}}$ (in units of $1/[2\pi(1-\nu)]$) by

$$\begin{aligned} \varepsilon_{xx}^{\mathrm{gr}} &= b_x y\left[(y^2 - \nu r^2)\Phi_1 + (3x^2 - y^2)\Phi_2\right] , \quad \varepsilon_{xz}^{\mathrm{gr}} = b_z(1-\nu) y r^2 \Phi_1/2 , \\ \varepsilon_{yy}^{\mathrm{gr}} &= b_x y\left[(x^2 - \nu r^2)\Phi_1 - (3x^2 - y^2)\Phi_2\right] , \quad \varepsilon_{yz}^{\mathrm{gr}} = -b_z(1-\nu) x r^2 \Phi_1/2 , \\ \varepsilon_{xy}^{\mathrm{gr}} &= -b_x x\left[y^2 \Phi_1 + (x^2 - 3y^2)\Phi_2\right] , \end{aligned} \tag{3.14}$$

with $\Phi_1 = K_1(r/\sqrt{c_2})/(\sqrt{c_2} r^3)$. For the stresses, the solution of (3.8) gives [151, 154] $\sigma_{ij} = \sigma_{ij}^0 + \sigma_{ij}^{\mathrm{gr}}$, where σ_{ij}^0 are given by (3.11) and $\sigma_{ij}^{\mathrm{gr}}$ (in units of $2D$) by

$$\begin{aligned} \sigma_{xx}^{\mathrm{gr}} &= \varepsilon_{xx}^{\mathrm{gr}}(\nu = 0, c_2 \to c_1) , \quad \sigma_{yy}^{\mathrm{gr}} = \varepsilon_{yy}^{\mathrm{gr}}(\nu = 0, c_2 \to c_1) , \\ \sigma_{zz}^{\mathrm{gr}} &= \nu(\sigma_{xx}^{\mathrm{gr}} + \sigma_{yy}^{\mathrm{gr}}) , \quad \sigma_{xy}^{\mathrm{gr}} = \varepsilon_{xy}^{\mathrm{gr}}(c_2 \to c_1) , \\ \sigma_{xz}^{\mathrm{gr}} &= \varepsilon_{xz}^{\mathrm{gr}}(c_2 \to c_1) , \quad \sigma_{yz}^{\mathrm{gr}} = \varepsilon_{yz}^{\mathrm{gr}}(c_2 \to c_1) . \end{aligned} \tag{3.15}$$

It is important to remember (see also Sect. 1.1) that the stress fields (3.15) for screw dislocations (when $b_x = 0$) are given by similar formulas in the nonlocal [141–143] and gauge field [159–161, 163, 164] theories. The solutions (3.14) and (3.15) for the screw and edge dislocations coincide with those in the gauge theory [163–165].

The main feature of the solution given by (3.13)–(3.15) is the absence of any singularities in the displacement, strain and stress fields. In fact, when $r \to 0$, we have $K_0(r/\sqrt{c_k})|_{r\to 0} \to -\gamma + \ln(2\sqrt{c_k}/r)$, where $\gamma = 0.5772\ldots$ is Euler's constant, $K_1(r/\sqrt{c_k}) \to \sqrt{c_k}/r$, $K_2(r/\sqrt{c_k}) \to 2c_k/r^2 - 1/2$ $(k = 1, 2)$. Hence, u_y is finite as $\varepsilon_{ij} \to 0$ and $\sigma_{ij} \to 0$. The fields of displacements (3.13) and strains (3.14) have been analysed in detail in [149, 150] using a special version of the gradient elasticity theory ($c_1 \equiv 0$). The stress fields (3.15) have been examined in [151]. We discuss below the main features of (3.13)–(3.15) separately for screw (Sect. 3.1.1.3) and edge (Sect. 3.1.1.4) dislocations.

Using (3.15), one can find the elastic energy of the dislocation within the gradient elasticity theory given by (3.5) as the work done by the gradient-dependent dislocation stress field (3.15) in producing the corresponding classical (for simplicity) plastic distortion $\beta^{*(\mathrm{cl})}_{yi} = (b_i/2)\delta(y)[1 - \mathrm{sign}(x)]$, $i = x, z$ [364]:

$$W_{\mathrm{s}} = -\frac{1}{2}\int_V \sigma_{zy}\beta^{*(\mathrm{cl})}_{yz}\mathrm{d}V$$

for screw dislocations and

$$W_{\mathrm{e}} = -\frac{1}{2}\int_V \sigma_{xy}\beta^{*(\mathrm{cl})}_{yx}\mathrm{d}V$$

for edge dislocations. The final result reads [151]

$$W = \frac{D}{2}\left\{\frac{b_x^2}{2} + \left[b_x^2 + (1-\nu)b_z^2\right]\left(\gamma + \ln\frac{R}{2\sqrt{c_1}}\right)\right\}. \tag{3.16}$$

We thus obtain a strain energy expression which is not singular at the dislocation line.

One can see [152] that the energy expression (3.16) contains only one gradient coefficient c_1, which looks strange. Recently, a more general consideration operating with the classical formula

$$W = \frac{1}{2}\int_V \sigma_{ij}\varepsilon_{ij}\mathrm{d}V$$

has resulted in the formula [365]

$$W = \frac{D}{2}\left\{\frac{3b_x^2}{4(1-\nu)} + \left[b_x^2 + (1-\nu)b_z^2\right]\left(\gamma + \ln\frac{R}{2\sqrt{c_1}} + \frac{c_2}{2(c_1 - c_2)}\ln\frac{c_2}{c_1}\right)\right\}, \tag{3.17}$$

which now contains both gradient coefficients c_1 and c_2.

3.1.1.3 Screw Dislocations

We now consider a screw dislocation whose line coincides with the z-axis of a Cartesian coordinate system (Fig. 3.1) and discuss briefly the main features [149, 151] of the gradient solution for dislocation fields (3.13)–(3.15) which distinguish it from the well-known classical solution (3.9)–(3.11).

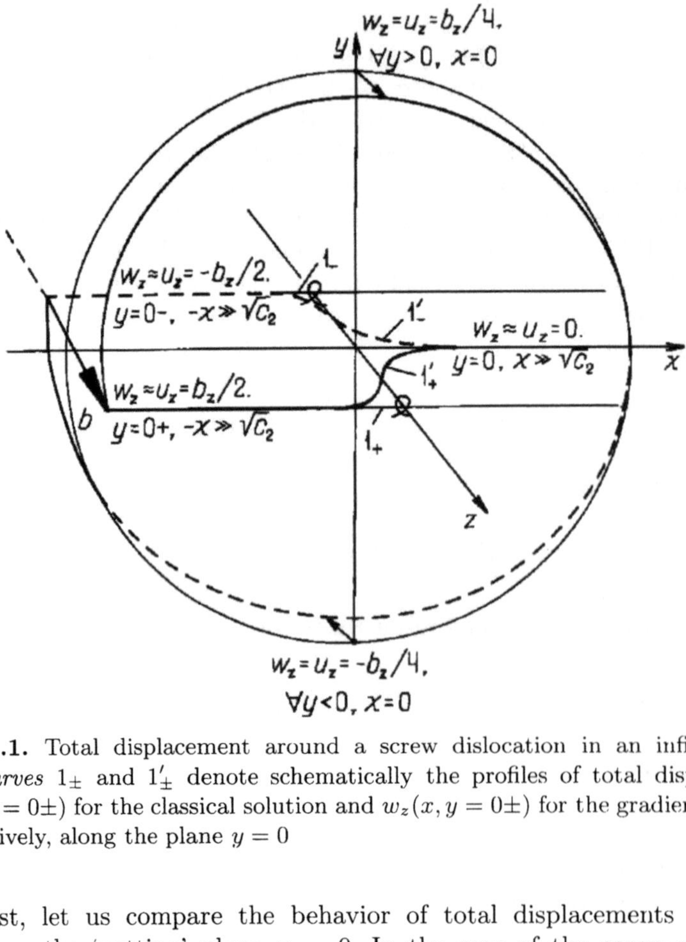

Fig. 3.1. Total displacement around a screw dislocation in an infinite body. The *curves* $1_\pm$ and $1'_\pm$ denote schematically the profiles of total displacements $u_z(x, y = 0\pm)$ for the classical solution and $w_z(x, y = 0\pm)$ for the gradient solution, respectively, along the plane $y = 0$

First, let us compare the behavior of total displacements (3.9) and (3.13) on the 'cutting' plane $y = 0$. In the case of the screw dislocation (Fig. 3.1), the displacement vector has only one non-vanishing component $w_z = (b_z/2\pi)w(x, y)$. When $y \to 0$, the integral in (3.13) may be evaluated in an explicit form [149] that gives

$$w(x, y \to 0) = \frac{\pi}{2}\text{sign}(y)\left[1 - \text{sign}(x)\left(1 - \mathrm{e}^{-|x|/\sqrt{c_2}}\right)\right] . \tag{3.18}$$

The additional term $-\mathrm{e}^{-|x|/\sqrt{c_2}}$ which appears in the gradient solution (3.18) leads to the smoothing of the total displacement profiles (curves $1'_+$ and $1'_-$

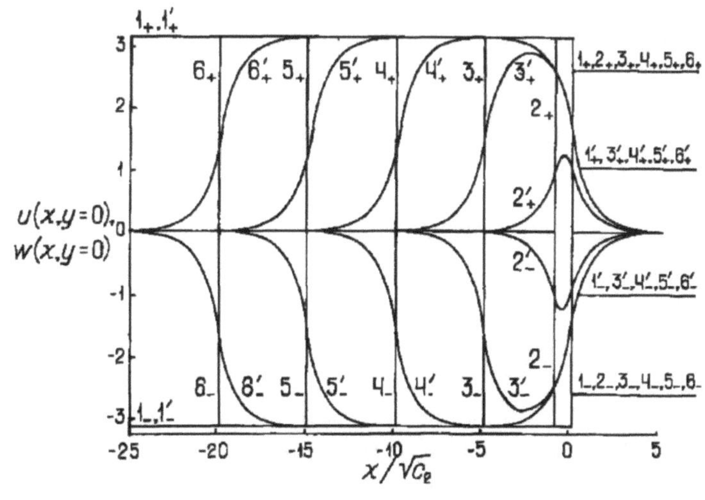

Fig. 3.2. Profiles of total displacements for a separate screw dislocation ($1_\pm$ for the classical solution and $1'_\pm$ for the gradient solution) and for a dislocation dipole with arm t ($2_\pm$–$6_\pm$ for the classical solution and $2'_\pm$–$6'_\pm$ for the gradient solution). The ratio $t/\sqrt{c_2}$ is taken as ∞ ($1_\pm$,$1'_\pm$), 1 ($2_\pm$,$2'_\pm$), 5 ($3_\pm$,$3'_\pm$), 10 ($4_\pm$,$4'_\pm$), 15 ($5_\pm$,$5'_\pm$), and 20 ($6_\pm$,$6'_\pm$)

in Figs. 3.1 and 3.2), in contrast to the abrupt jumps occurring in these profiles in the classical solution (curves 1_+ and 1_- in Figs. 3.1 and 3.2). It is interesting to note that the size of such a transition zone is approximately $10\sqrt{c_2}$, which gives the value $2.5a$ for a crystalline lattice, i.e., the usual size of the dislocation core. It follows that, in gradient elasticity, the dislocation core appears naturally as the result of a calculation defining the region of maximum strain around the dislocation line [149]. In classical elasticity, there is no definition for the dislocation core and the dislocation is treated as a linear singularity of elastic stress and strain fields.

Second, let us consider now a gradient solution for a dipole of screw dislocations [149]. As usually, such a solution may be found simply by superposing solutions for separate dislocations. The only reason for discussing this case here is the appearance of a new interesting effect concerning the behavior of the total displacement of a dipole. Consider two parallel screw dislocations lying in the plane $y = 0$ along the z-axis and crossing the x-axis at the points $x = -t$ and $x = 0$. Let us assume the same Burgers vector b_z and opposite tangent vectors $\pm l_z$. This means that they can be treated as a limiting case of a rectangular gliding dislocation loop with edge segments at $z = \pm\infty$. In the framework of classical dislocation theory, the field of total displacement for such a dislocation configuration is described by the vector $u_z = (b_z/2\pi)u(x,y)$, where

$$u(x,y) = \arctan\frac{y}{x} - \arctan\frac{y}{x+t} + \frac{\pi}{2}\text{sign}(y)\big[\text{sign}(x+t) - \text{sign}(x)\big] \ . \quad (3.19)$$

The gradient solution is described by the vector $w_z = (b_z/2\pi)w(x,y)$, where

$$w(x,y) = u(x,y) - \mathrm{sign}(y)\frac{\mathrm{i}}{2}\int_{-\infty}^{+\infty}\left(1-\mathrm{e}^{\mathrm{i}ts}\right)\frac{s}{1/c_2+s^2}\exp\left(-|y|\sqrt{\frac{1}{c_2}+s^2}\right)\mathrm{e}^{\mathrm{i}sx}\mathrm{d}s\,, \tag{3.20}$$

and $u(x,y)$ is given by (3.19). When $y \to 0$,

$$w(x,y\to 0) \tag{3.21}$$
$$= \frac{\pi}{2}\mathrm{sign}(y)\left[\mathrm{sign}(x+t)\left(1-\mathrm{e}^{-|x+t|/\sqrt{c_2}}\right) - \mathrm{sign}(x)\left(1-\mathrm{e}^{-|x|/\sqrt{c_2}}\right)\right].$$

The graphs for $u(x,y\to 0)$ and $w(x,y\to 0)$ are presented in Fig. 3.2. Besides the evident difference in the form of these profiles, it is important to note that the maximum value of the total displacement depends on the dipole arm t for the gradient solution w, in contrast to the classical solution u which is independent of t. This dependence disappears when $t \geq t_0 \approx 10\sqrt{c_2}$, where t_0 defines a new characteristic distance, namely, the radius of the strong short-range interaction between dislocations. For a crystalline lattice, this gives $t_0 \approx 2a$, a result consistent with intuition.

As pointed out in Sect. 3.1.1.2, the main feature of the gradient solution given by (3.14) and (3.15) is the absence of any singularities in either the strain or the stress fields. To illustrate the characteristic features of the gradient solution, the spatial distribution of elastic strains ε_{iz} and stresses σ_{iz} ($i = x,y$) near the dislocation line [151, 152] is presented in Fig. 3.3. One can see that the gradient solutions for the elastic strains and stresses attain their extreme values of approximately $\pm b_z/(10\pi\sqrt{c_2})$ and $\pm Gb_z/(5\pi\sqrt{c_1})$ at distances $\approx \sqrt{c_2}$ and $\sqrt{c_1}$ from the dislocation line, respectively. Using for the gradient coefficients c_1 and c_2 the estimate [355, 356] $\sqrt{c_1} \approx \sqrt{c_2} \approx a/4$ for a crystalline lattice, we find that $\max|\varepsilon_{iz}| \approx 12\%$ [149] and $\max|\sigma_{iz}| \approx G/4$ [151] at a distance $\approx a/4$ from the dislocation line. It is also seen that the gradient solutions coincide with the classical ones far away from the dislocation core ($r \geq r_0 \approx 4\sqrt{c_k}$) [149].

The elimination of singularities from the strain and stress fields allows us to consider in detail the short-range nanoscale interaction between dislocations. It was shown above that this short-range interaction takes place when the spacing t between dislocations is smaller than $\approx 10\sqrt{c_2} \approx 2.5a$. Here we consider two simple dislocation configurations: a dipole and a pair of screw dislocations (Fig. 3.4) [152]. The elastic fields of these configurations are obtained as simple superpositions of the corresponding fields for individual dislocations. Figure 3.5 shows the distribution of the strain and stress components $\varepsilon_{yz}(x,0)$ and $\sigma_{yz}(x,0)$ near the dislocation dipole (Fig. 3.5a and b) and dislocation pair (Fig. 3.5c and d). One can see that the strain and stress are finite at the dislocation lines and tend to zero there when the magnitude of the interdislocation spacing d increases. Between dislocations or near

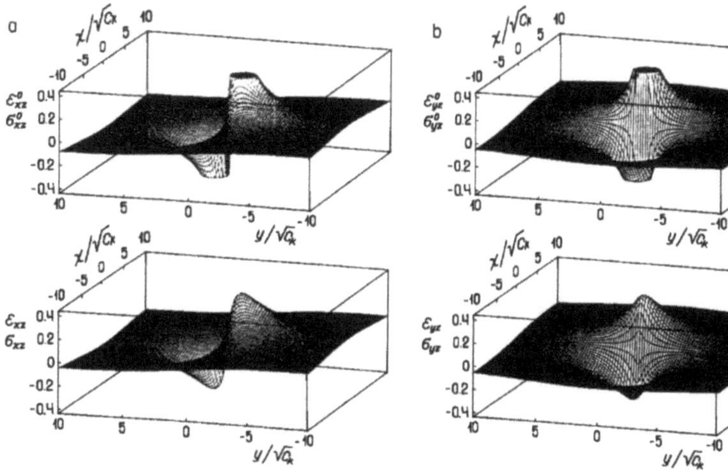

Fig. 3.3. Coordinate-dependent parts of the strain ε^0_{xz}, ε_{xz} (**a**), ε^0_{yz}, ε_{yz} (**b**), or stress σ^0_{xz}, σ_{xz} (**a**), σ^0_{yz}, σ_{yz} (**b**) components near the line of a screw dislocation which goes through the point $(0,0)$. Strain values are given in units of $b_z/(4\pi\sqrt{c_2})$ and stress values in units of $Gb_z/(2\pi\sqrt{c_1})$. The *upper figures* give the classical solutions ε^0_{iz}, σ^0_{iz} and the *lower figures* give the gradient solution ε_{iz}, σ_{iz} $(i = x, y)$

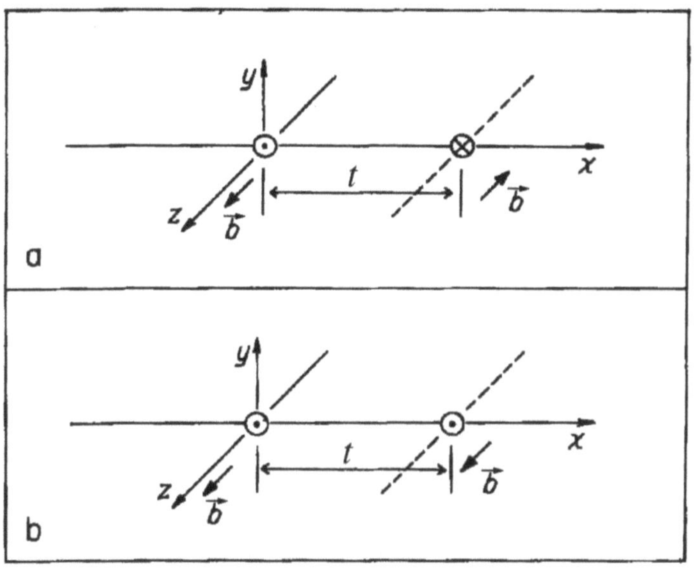

Fig. 3.4. The dipole (**a**) and pair (**b**) of screw dislocations

them, where the classical solutions (dashed curves) give unreasonably high strain and stress values, the gradient solutions (continuous curves) give more reasonable values, $\varepsilon_{yz} \leq 25\%$ and $\sigma_{yz} \leq G/2$.

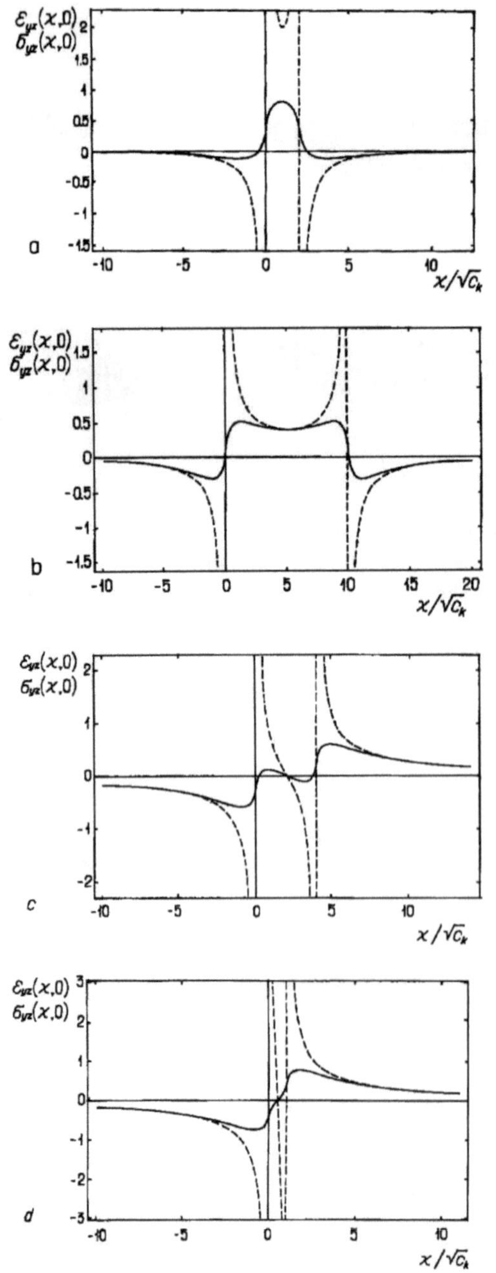

Fig. 3.5. Coordinate-dependent parts of strain $\varepsilon_{yz}(x, 0)$ or stress $\sigma_{yz}(x, 0)$ near the dipole (**a**, **b**) and pair (**c**, **d**) of screw dislocations for different values of the interdislocation spacing $t/\sqrt{c_k} = 2$ (**a**), 10 (**b**), 4 (**c**), and 1 (**d**), where $k = 2$ corresponds to the strain distribution and $k = 1$ to the stress distribution. The strain values are given in units of $b_z/(4\pi\sqrt{c_2})$ and the stress values in units of $Gb_z/(2\pi\sqrt{c_1})$. *Dashed curves* represent the classical solution

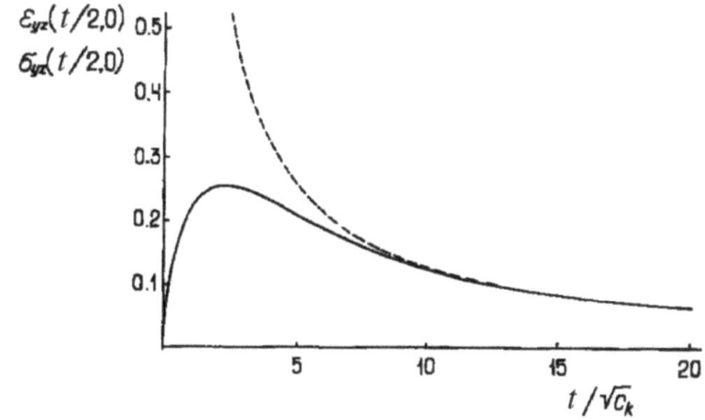

Fig. 3.6. Strain and stress components, $\varepsilon_{yz}(t/2, 0)$ and $\sigma_{yz}(t/2, 0)/(2G)$, in the middle of the dipole of screw dislocations via the dipole arm $t/\sqrt{c_k}$ for the case $b_z = 4\sqrt{c_k}$, where $k = 2$ corresponds to strain distribution and $k = 1$ to stress distribution. The *dashed curve* represents the classical solution

In the case of dislocation dipoles, the strain and stress at the central point between the dislocations do not exceed these levels and tend to zero when $t \to 0$ (Fig. 3.6). In the case of a dislocation pair, the levels of strain and stress decrease between the dislocation lines as they glide towards each other and remain finite all the way up to the point where the two dislocations come together at the same line. One can conclude [152] that the short-range elastic interaction between dislocations is much smaller than that predicted from classical elasticity. This means that there are no significant energetic barriers for the processes of elementary nucleation of dislocation dipoles or formation of superdislocations in high-density dislocation ensembles.

3.1.1.4 Edge Dislocations

Consider an edge dislocation whose line coincides with the z-axis while the Burgers vector $\boldsymbol{b} = b_x \boldsymbol{e}_x$ is parallel to the x-axis of the Cartesian coordinate system. The most interesting question here is the behavior of the total displacement and elastic strain and stress near the dislocation line [150, 151].

First, consider the components u_x and u_y of the total displacement given by (3.13) on the 'cutting plane' $y = 0$. When $y \to 0$, we have [150]

$$u_x(x, y \to 0) = \frac{b_x}{4}\mathrm{sign}(y)\left[1 - \mathrm{sign}(x)\left(1 - \mathrm{e}^{-|x|/\sqrt{c_2}}\right)\right] , \tag{3.22}$$

$$u_y(x, y \to 0) = \frac{b_x}{4\pi(1-\nu)}\left[-1 - (1-2\nu)\ln|x| + \frac{2c_2}{x^2} - (1-2\nu)K_0\left(\frac{|x|}{\sqrt{c_2}}\right) - K_2\left(\frac{|x|}{\sqrt{c_2}}\right)\right] . \tag{3.23}$$

Equation (3.22) coincides with (3.18), which describes the total displacement of a screw dislocation at $y = 0$. The term $-e^{-|x|/\sqrt{c_2}}$ which appears in the gradient solution (3.22) leads to the smoothing of the total displacement profile in contrast to the abrupt jump occurring in this profile in the classical solution (see Figs. 3.1 and 3.2). This means that in gradient elasticity, the dislocation core appears again naturally, directly from the calculations, as discussed above for the screw dislocation.

Equation (3.23) contains the terms which are singular at the dislocation line. However, they compensate each other. To demonstrate this correctly, let us consider here the field of total displacements created by a dipole of edge dislocations, thus avoiding the logarithm of a dimensional quantity.

Let two parallel edge dislocations lie in the plane $y = 0$ along the z-axis and cross the x-axis at the points $x = -t$ and $x = 0$. Let us assume the same Burgers vector b_x and opposite tangent vectors $\pm l_z$. This means that they can be treated as a limiting case of a rectangular gliding dislocation loop with screw segments at $z = \pm\infty$. Both classical and gradient solutions may be obtained by a simple superposition of the solutions for a separate dislocation. In particular, for $y \to 0$, we have [150]

$$u_x|_{y\to 0} = \frac{b_x}{2}\mathrm{sign}(y)\left[\mathrm{sign}(x+t)\left(1 - e^{-|x+t|/\sqrt{c_2}}\right) - \mathrm{sign}(x)\left(1 - e^{-|x|/\sqrt{c_2}}\right)\right], \tag{3.24}$$

$$\begin{aligned} u_y|_{y\to 0} = \frac{b_x}{4\pi(1-\nu)}\Bigg\{&-(1-2\nu)\left[\ln\frac{|x|}{|x+t|} + K_0\left(\frac{|x|}{\sqrt{c_2}}\right) - K_0\left(\frac{|x+t|}{\sqrt{c_2}}\right)\right] \\ &+ \frac{2c_2}{x^2} - \frac{2c_2}{(x+t)^2} - K_2\left(\frac{|x|}{\sqrt{c_2}}\right) + K_2\left(\frac{|x+t|}{\sqrt{c_2}}\right)\Bigg\}. \end{aligned} \tag{3.25}$$

The component $u_x(y \to 0)$ given by (3.24) coincides with the component $w_z(y \to 0)$ of a similar dipole of screw dislocations [see (3.21) and Figs. 3.1 and 3.2). From this analysis, one can conclude that in gradient elasticity a new characteristic distance appears, namely the radius of strong short-range interaction between dislocations. It follows from the identity of (3.21) and (3.24) that this conclusion is also valid for the case of edge dislocations [150].

The component $u_y(y \to 0)$ given by (3.25) is regular everywhere including the dislocation lines. In fact, for $x \to 0$, we have

$$K_0(|x|/\sqrt{c_2})|_{x\to 0} \to -\gamma - \ln\frac{|x|}{2\sqrt{c_2}} \quad \text{and} \quad K_2(|x|/\sqrt{c_2})|_{x\to 0} \to 2c_2/x^2 - 1/2\,.$$

Hence,

$$\begin{aligned} u_y(0,0) = \frac{b_x}{4\pi(1-\nu)}\Bigg\{&(1-2\nu)\left[\gamma + \ln\frac{t}{2\sqrt{c_2}} + K_0\left(\frac{t}{\sqrt{c_2}}\right)\right] + \frac{1}{2} \\ &-\frac{2c_2}{t^2} + K_2\left(\frac{t}{\sqrt{c_2}}\right)\Bigg\}. \end{aligned} \tag{3.26}$$

It can be seen from (3.26) that $u_y(0,0)$ increases with the dipole arm t and equals zero when $t = 0$. The dependence of $u_y(x, y = 0)$ on t is presented in Fig. 3.7, which demonstrates the fact that the gradient solution is especially effective for describing the strong short-range interaction between dislocations.

The plots of Fig. 3.7c were obtained numerically in [150] for a dislocation dipole with arm $t = 100\sqrt{c_2}$, which may thus approximate the field of a separate dislocation. The gradient solution gives smaller displacements near the dislocation line ($-0.3\sqrt{c_2} \le x \le 0.3\sqrt{c_2}$) and larger ones outside of this region. With these plots, we can compare the calculated value of the displacement $u_y(0,0)$ at the dislocation line with data from experimental observations and computer simulations. Considering u_y for a dislocation dipole with arm $t = 100\sqrt{c_2}$, we have the estimate $u_y(0,0) \approx 2.3b_x/[4\pi(1-\nu)]$, which gives $\approx 0.26a$ for $b_x = a$ and $\nu = 0.3$. This value agrees well with the results of direct observation of edge dislocations near an Nb/Al_2O_3 interface [366], as well as with the results of computer simulations for the core of an edge dislocation in α-Fe [367].

Let us now consider the behavior of the elastic strain and stress components given by superpositions of (3.10)–(3.11) with (3.14)–(3.15) near the dislocation line. The main feature of the gradient solution is the absence of any singularities in the strain and stress fields. One can observe some interesting details in the behavior of this solution [150].

From Fig. 3.8, one can conclude that the component $\varepsilon_{xx}(x=0)$ is smaller than $\varepsilon^0_{xx}(x=0)$ for $0 \le |y| \le 1.3\sqrt{c_2}$ and larger than $\varepsilon^0_{xx}(x=0)$ for $|y| \ge 1.3\sqrt{c_2}$. It reaches a maximum value ≈ 0.3 (all strain values are given here in units of $b_x/[4\pi(1-\nu)\sqrt{c_2}]$) at $|y| \approx \sqrt{c_2}$, which gives $\approx$14% for an atomic lattice. The component $\varepsilon_{yy}(x=0)$ is smaller than $\varepsilon^0_{yy}(x=0)$ everywhere. It is equal to zero at $|y| \approx 0.6\sqrt{c}$ and achieves two extremum values of opposite signs (on the same side of the dislocation line), equal to ≈ 0.01 ($\approx 0.5\%$) at $\approx 0.2\sqrt{c_2}$, and ≈ 0.06 ($\approx 3\%$) at $\approx 4\sqrt{c_2}$. It is interesting to note that $\varepsilon_{yy}(x=0)$ is significantly smaller than $\varepsilon_{xx}(x=0)$, in contrast to the classical solution where $\varepsilon^0_{yy}(x=0) \equiv \varepsilon^0_{xx}(x=0)$. The behavior of the dilatation $\varepsilon = \varepsilon_{xx} + \varepsilon_{yy}$ is similar to the behavior of the strain component ε_{xz} of a screw dislocation (see Sect. 3.1.1.3). The dilatation has a maximum value of about 0.3 ($\approx 14\%$) at $|y| \approx \sqrt{c_2}$. It is equal to zero at the dislocation line and practically coincides with the classical solution outside of the dislocation core ($|y| > 4\sqrt{c_2}$). The shear component ε_{xy} is smaller than ε^0_{xy} everywhere. It approaches zero at the dislocation line, achieves maximum values of about 0.25 ($\approx 12\%$) at $|x| \approx 1.5\sqrt{c_2}$, and practically coincides with ε^0_{xy} far from the dislocation line ($|x| > 10\sqrt{c_2}$).

A general (three-dimensional) view of the spatial distribution of elastic strains near the dislocation line [152] is provided in Fig. 3.9, which demonstrates all the aforementioned features of the gradient solution for the strain field.

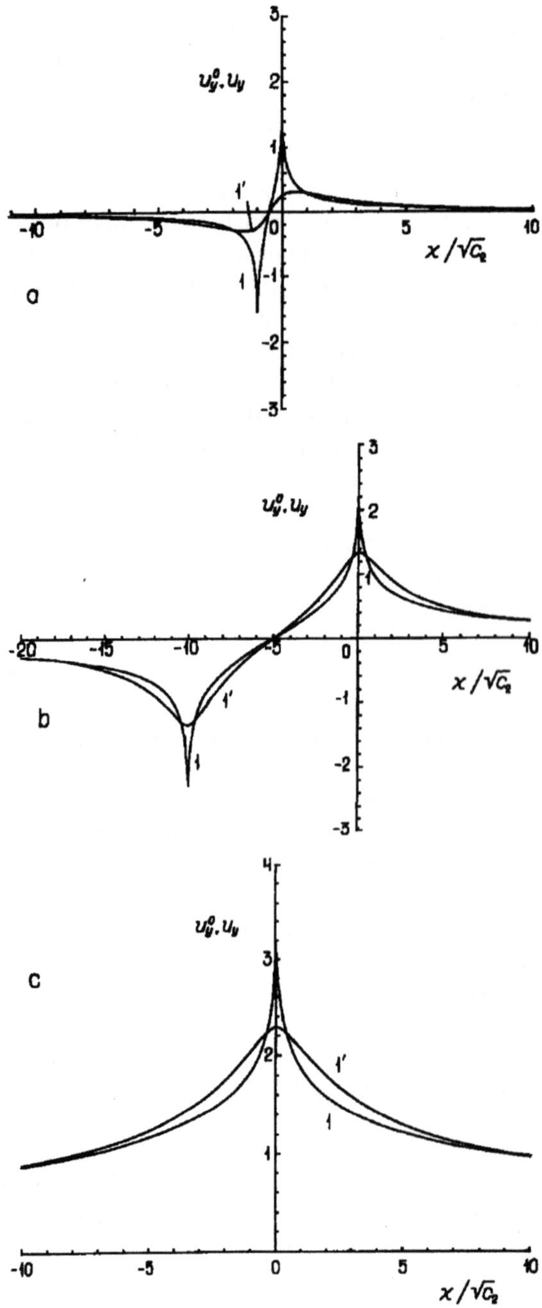

Fig. 3.7. y-components of total displacement of a dipole of edge dislocations along the plane $y = 0$ when the dipole arm t is equal to $1\sqrt{c_2}$ (**a**), $10\sqrt{c_2}$ (**b**) and $100\sqrt{c_2}$ (**c**). (1) Classical solution $u_y^0(x, y = 0)$, (1′) gradient solution $u_y(x, y = 0)$. Displacement values are given in units of $b_x/[4\pi(1-\nu)]$

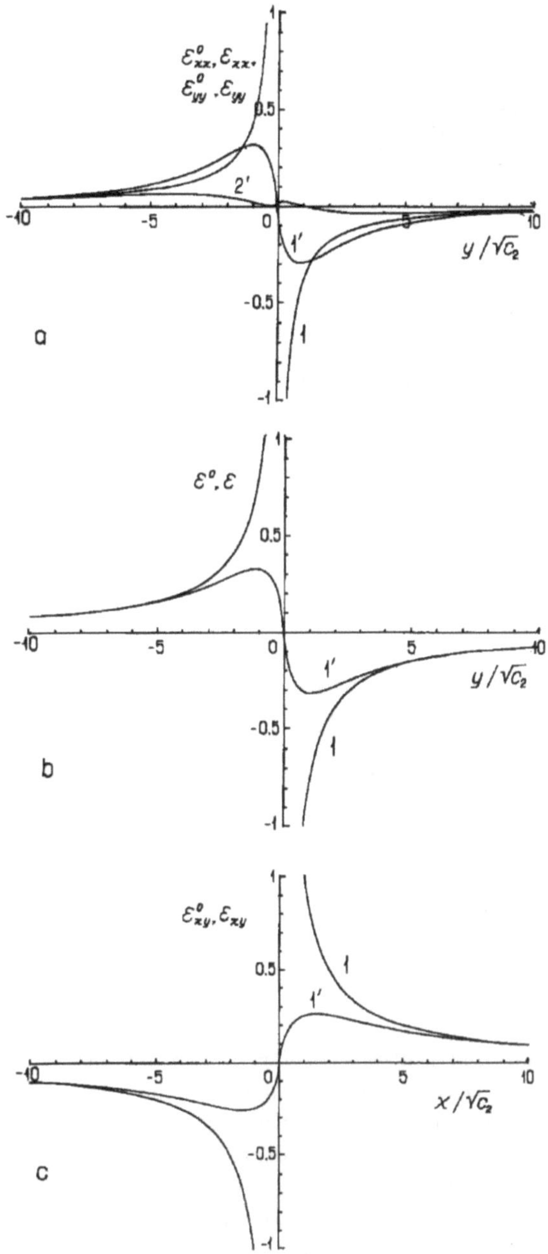

Fig. 3.8. Strain components (**a**) $\varepsilon^0_{xx}(x=0,y) \equiv \varepsilon^0_{yy}(x=0,y)$ (curve 1), $\varepsilon_{xx}(x=0,y)$ (curve 1′), $\varepsilon_{yy}(x=0,y)$ (curve 2′), (**b**) $\varepsilon^0(x=0,y)$ (curve 1), $\varepsilon(x=0,y)$ (curve 1′), (**c**) $\varepsilon^0_{xy}(x,y=0)$ (curve 1), $\varepsilon_{xy}(x,y=0)$ (curve 1′) near the dislocation line. Strain values are given in units of $b_x/[4\pi(1-\nu)\sqrt{c_2}]$

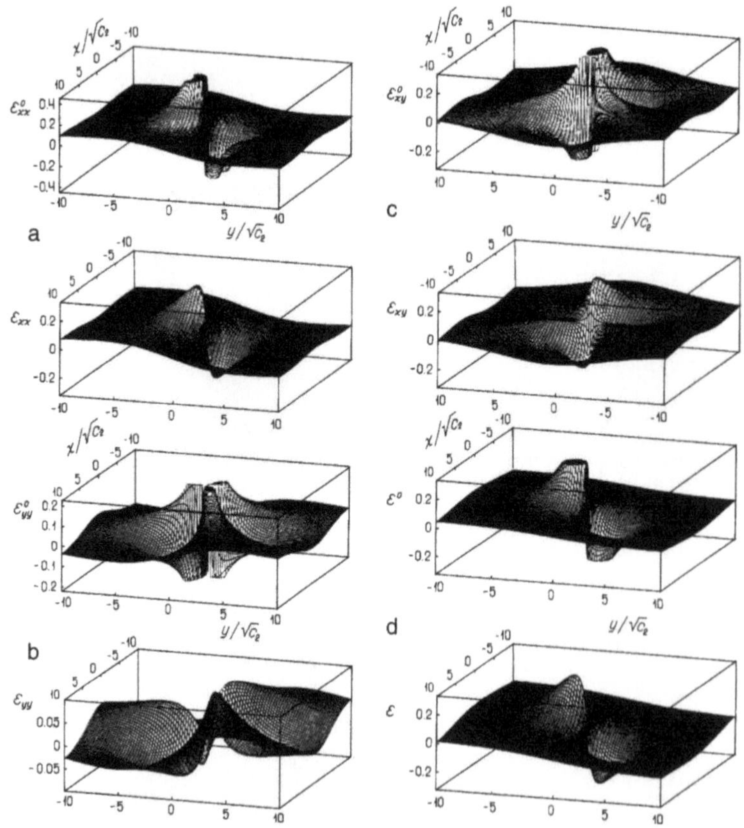

Fig. 3.9. Strain components ε_{xx} (**a**), ε_{yy} (**b**), ε_{xy} (**c**), and ε (**d**) near the line of an edge dislocation which goes through the point $(0, 0)$. Strain values are given in units of $b_x/[4\pi(1-\nu)\sqrt{c_2}]$. The *upper figures* give the classical singular solutions while the *lower ones* give the gradient regular solutions

Similar spatial distributions of stress components σ_{xx} and σ_{yy} near the dislocation line [151] are represented in Fig. 3.10. The distribution of the shear stress component σ_{xy} is the same as that of the shear strain component ε_{xy} (Fig. 3.9c) with appropriate replacement of strain units $b_x/[4\pi(1-\nu)\sqrt{c_2}]$ by stress units $Db_x/\sqrt{c_1}$ and coordinate units $(x, y)/\sqrt{c_2}$ by $(x, y)/\sqrt{c_1}$. Moreover, the distribution of the hydrostatic stress component $\sigma = (\sigma_{xx}+\sigma_{yy}+\sigma_{zz})/3$ is the same as that of the dilatation ε (Fig. 3.9d) with similar replacement of units $b_x/[4\pi(1-\nu)\sqrt{c_2}]$ by $Db_x(1+\nu)/[3(1-2\nu)\sqrt{c_1}]$, and $(x, y)/\sqrt{c_2}$ by $(x, y)/\sqrt{c_1}$. One can see that the gradient solutions for the stresses attain their extreme values $|\sigma_{xx}| \approx 0.45G$ and $|\sigma_{yy}| \approx |\sigma_{xy}| \approx 0.27G$ for $b_x = a = 4\sqrt{c_1}$ and $\nu = 0.3$ at a distance $\approx a/4$ from the dislocation line. It can also be seen that the gradient solutions coincide with the classical ones far away from the dislocation core ($r \geq r_0 \approx 4\sqrt{c_k}$) [150].

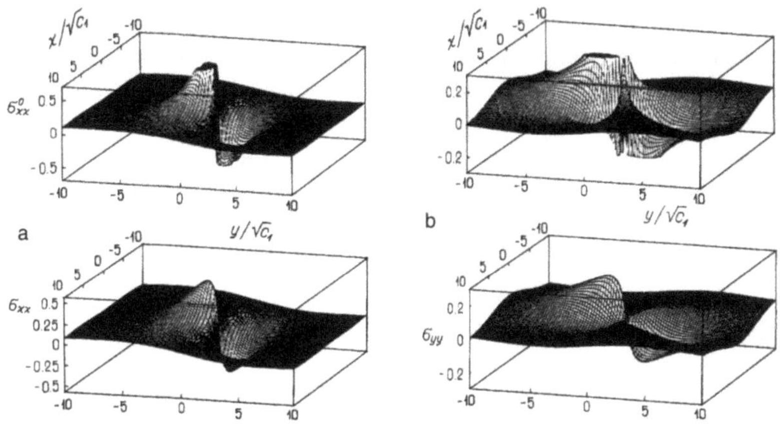

Fig. 3.10. Stress components σ^0_{xx}, σ_{xx} (**a**) and σ^0_{yy}, σ_{yy} (**b**) near the line of an edge dislocation which goes through the point $(0, 0)$. Stress values are given in units of $Db_x/\sqrt{c_1}$. The *upper figures* give the classical solutions σ^0_{ij} while the *lower ones* give the gradient solution σ_{ij}

Considering the straight dislocations in a homogeneous elastic isotropic solid within the gradient theory of elasticity described by (3.5), one can therefore sum up by saying that the new gradient solution gives non-singular expressions for the dislocation displacement, strain and stress fields and elastic energies. It has been shown that, for individual dislocations, the elastic strains and stresses are strictly equal to zero at the dislocation lines and achieve their extreme values of $\approx$ (3–14)% and $\approx G/4$–$G/2$, respectively, at a distance $\approx a/4$ from the dislocation line. Two characteristic distances appear naturally in this approach: $r_0 \approx 4\sqrt{c_2}$, which may be viewed as the radius of the dislocation core and $t_0 \approx 10\sqrt{c_2}$ which may be viewed as the radius of the strong short-range interaction between dislocations.

After such a long discussion of non-singular dislocation fields obtained within gradient elasticity, let us come back to cellular dislocations. Now one can apply all the aforementioned advantages of the gradient solutions to describing the elastic fields of cellular dislocations. To construct the gradient model for a cellular dislocation [351], it is enough to make the substitutions $b_i = B_i \approx d$, where $i = x, z$, and $4\sqrt{c_i} \approx d$. This leads to a cellular dislocation which has a core size and Burgers vector value of about the grain size d. One can then say that Figs. 3.1–3.10 give an impression of the distribution of displacements, strains and stresses near the lines of cellular dislocations.

Summing up, the suggested gradient model of a cellular dislocation allows one to avoid the main difficulties in the description of cellular dislocations, occurring within the framework of the classical theory of defects, i.e., one can avoid abnormally high displacements, strains and stresses within the cores of cellular dislocations, and one can also characterize the cellular dislocations by well known classical long-range elastic fields far from the core regions.

We need these long-range fields to describe long-range interactions between cellular dislocations, while knowledge of the strained state within their cores is needed to model the cellular dislocation motion. As we have seen above, the values of elastic stresses within the dislocation core remain very high, of the order of the theoretical strength, even in the gradient elasticity theory. This means that some relaxation processes are necessary in the cores of cellular dislocations. This question needs additional experimental and theoretical investigation, and the present gradient model seems to be a good basis for it.

3.1.2 Nucleation and Motion of Cellular Dislocations

Following [351], let us apply the concept of cellular dislocations to a description of plastic flow in nanostructured ceramic Ti–B–N films studied experimentally by Andrievskii et al. [324, 325].

Consider the situation near the indenter tip. High contact elastic strains and stresses cause grain boundary dislocations to glide along the grain boundary surfaces which coincide with the planes of maximal shear stress. It is assumed that the average size of nanograins is small enough to have many joined grain boundaries which lie in the same plane of maximal shear stress, thus forming the grain boundary sliding surface. Gliding along such surfaces, grain boundary dislocations meet various obstacles, such as small misorientations of joined grain boundaries, and accumulate nearby, awaiting further motion of the indenter tip. When the indenter penetrates more deeply, it creates additional impulses of contact stresses and makes grain boundary dislocations pursue their gliding motion past obstacles. As a result, these accumulations of grain boundary dislocations are transformed into new entities, treated here as cellular dislocations. It is worth noting that such a transformation may include elementary coupling processes between grain boundary dislocations which are very hard to describe on the basis of classical considerations of dislocations due to the infinite forces of interdislocation interaction at a distance smaller than the core radius. However, in the framework of gradient elasticity, it is possible to analyse such short-range interactions, avoiding these difficulties (see Sects. 3.1.1.3 and 3.1.1.4). We can thus imagine cellular dislocations as high density agglomerations of $N = B/b$ grain boundary dislocations, where b is the Burgers vector of the latter.

During their motion, cellular dislocations stimulate new grain boundary dislocations to glide and involve them in cooperative gliding, thereby increasing their elastic fields and energy. On the other hand, part of their energy is expended on various accommodation processes for grain boundary sliding. In this context, the accommodation processes may control the motion of cellular dislocations and, therefore, peculiarities of plastic deformation. It was thus assumed [351] that plastic flow in a nanostructured film under the action of an indenter is either homogeneous or inhomogeneous, depending on the contribution of accommodation processes to the plastic deformation.

When the accommodation processes are not intensive, one can expect the cellular dislocations to move at low values of the flow stress. In this case, plastic relaxation will occur through the generation and short-path gliding of cellular dislocations along many available grain boundary sliding surfaces which lie close to the planes of maximal shear stress. This corresponds to the homogeneous regime of plastic flow, and one can expect this scenario in NCMs with bigger grain size.

When the accommodation processes are very intensive, one can expect the cellular dislocations to move at high values of the flow stress. In this case, the contact elastic fields created by the indenter will relax through the action of some limited number of grain boundary sliding surfaces which all lie close to each other, along the planes of maximal shear stresses, thereby forming shear bands. This corresponds to the inhomogeneous regime of plastic flow. It is natural to expect this scenario in NCMs with smaller grain size, where the volume fraction of grain boundaries achieves high values and there are many more possibilities for finding planar grain boundary sliding surfaces.

To illustrate the above statements, let us consider the two following situations with respectively low and high contributions of accommodation processes to the plastic deformation occurring via the motion of cellular dislocations.

1. Plastic deformation occurs via the motion of cellular dislocations, involving only low-intensity accommodation processes. In such a situation, the cellular dislocations move in a deformed film at low values of the yield stress, since the accommodation weakly contributes to the hampering force which acts on the moving cellular dislocations. As a corollary, if a cellular dislocation is generated, it immediately starts moving in a deformed film. However, local structural and chemical inhomogeneities induce local increases in the hampering force which acts on moving cellular dislocations in some local regions of the film. In other words, the accommodation processes accompanying motion of the cellular dislocations can be highly intensive in some local regions of the film due to its structural or chemical inhomogeneities. This can result in stopping motion of a cellular dislocation in such a local region, in which case the yield stress (slightly) increases. Increased yield stress generates a new cellular dislocation in another local region of the deformed film, which immediately starts moving. In this way, cellular dislocations are generated and move in a spatially uncorrelated manner in different regions of the film, providing its homogeneous deformation.
2. Plastic deformation occurs via the motion of cellular dislocations, involving high-intensity accommodation processes. In such a situation, the cellular dislocations move in a deformed film only at high values of the yield stress, since the accommodation essentially contributes to the hampering force which acts on the moving cellular dislocations. As a corollary, at the first stage of deformation, many cellular dislocations are generated in

the film which are immobile at low and intermediate values of the yield stress. In other words, the first stage of deformation occurring via the generation of numerous cellular dislocations leads to generation and accumulation of the dislocation charge associated with the immobile cellular dislocations, approximately homogeneously distributed in the film. The second stage of deformation comes into play when a previously stopped cellular dislocation starts moving under the action of a high yield stress. This cellular dislocation approaches its neighbouring cellular dislocation, in which case a pile-up of the two cellular dislocations is formed. Due to the effect of stress concentration on the head cellular dislocations of the pile-up, it proceeds to move, involving new cellular dislocations in its motion and resulting in the formation of a shear band characterized by a high degree of plastic deformation. After the first shear band has propagated through a section of the deformed film, a new band is formed and evolves in another local region of the film. This scenario is repeated, thus providing inhomogeneous deformation of the film with superplastic flow, localized in narrow shear bands.

We have thus considered how accommodation processes can cause either homogeneous or inhomogeneous deformation regimes. In doing so, the theoretical scheme considered is independent of any concrete mechanism for the accommodation processes. Let us now discuss, with the help of the scheme [351], the specific features of homogeneous and inhomogeneous regimes of plastic deformation in nanostructured films.

As has been shown in experiments [324, 325], the plastic flow in TiB_2 nanostructured films with hexagonal structure (AlB_2-type) is inhomogeneous, being localized in shear bands, while TiN nanostructured films with cubic structure (NaCl-type) are homogeneously deformed. In [325], it is proposed that the difference in plastic behavior between films with the hexagonal and cubic structure is related to the difference in anisotropy and slip systems. Hence, primary slip systems for the hexagonal TiB_2 are only $\{1010\}$ $\langle 1120\rangle$ and $\{1010\}$ $\langle 0001\rangle$, while the face-centered-cubic TiN has $\{111\}$ $\langle 110\rangle$ polyslip. The proposal [325] serves as a natural constituent of the theoretical scheme [351] in the specific case discussed, if the accommodation processes for the motion of cellular dislocations in nanostructured boride and nitride films are associated with dislocation slip in nanograin interiors. In this context, the accommodation dislocation slip in nanograin interiors in films with the cubic structure is facilitated (due to polyslip). This results in situation (1) above, i.e., the situation with low-intensity accommodation processes which cause the homogeneous deformation. Accommodation dislocation slip in nanograin interiors in films with the hexagonal structure is hampered (due to slip in only two slip planes) resulting in situation (2) above, i.e., the situation with high-intensity accommodation processes which cause the inhomogeneous deformation. In any event, however, new (detailed) experimental studies are of interest if they can identify the deformation micromechanism(s) in nanostructured boride and nitride films and therefore either support or refute the theoretical scheme suggested in [351].

3.1.3 Kinetics of Cellular Dislocations

On the basis of the aforementioned mechanisms for nucleation and motion of cellular dislocations, we have proposed a model of evolutional kinetics for their ensemble [368]. The model leads to a system of coupled chemical-reaction-kinetics differential equations with respect to the densities of mobile and immobile cellular dislocations, which has an exact solution given in an explicit form. The solution may further be used to model the behavior of cellular dislocations in a wide range of experimental conditions.

Consider an ensemble of unidirectional cellular dislocations, one part of which is assumed to be mobile while the other part is immobile. Let the densities of the mobile and immobile cellular dislocations be ρ and n, respectively. We suppose that the nucleation, motion and immobilization of cellular dislocations occur locally within narrow shear bands which lie in the xz-plane of a rectangular coordinate system. The x-axis shows the direction of gliding for cellular dislocations and the z-axis is parallel to their lines.

The following scenario of evolution of the cellular dislocation ensemble is assumed to be realized. In the first stages, the usual lattice and grain boundary dislocations accumulate near some obstacles and create immobile cellular dislocations (n-dislocations). Let their generation rate be D. As the external load increases, the weaker obstacles are destroyed and the corresponding n-dislocations pass through them, thus being transformed into mobile cellular dislocations (ρ-dislocations). The frequency of such a transformation is denoted by B. The ρ-dislocations glide with mean velocity A and, meeting n-dislocations, help them to pass through stronger obstacles and also be transformed into ρ-dislocations. The rate of this process is characterized by the parameter C.

The evolution of the model ensemble of ρ- and n-dislocations may be described by the following system of two rate equations [368]:

$$\frac{\partial \rho}{\partial t} + A\frac{\partial \rho}{\partial x} = Bn + Cn\rho \,, \tag{3.27}$$

$$\frac{\partial n}{\partial t} = D - Bn - Cn\rho \,, \tag{3.28}$$

where t denotes the time. To solve the system of evolution equations (3.27)–(3.28), let us pass from the immobile (x, y, z) coordinate system which is connected with the sample to a mobile (s, y, z) coordinate system which moves along the x-direction with velocity v, i.e., $s = x - vt$. In this mobile coordinate system, (3.27)–(3.28) are given by

$$(A - v)\frac{\partial \rho}{\partial s} = Bn + Cn\rho \,, \tag{3.29}$$

$$v\frac{\partial n}{\partial s} = -D + Bn + Cn\rho\,. \tag{3.30}$$

By subtracting (3.30) from (3.29) and integrating the difference, one can find a simple relationship between the densities ρ and n, viz.,

$$(A-v)\rho = vn + Ds + C_1\,, \tag{3.31}$$

where C_1 is a constant.

Putting (3.31) into (3.30) gives the Riccati equation [369] for n which reads

$$n'_s = an^2 + (bs + c)n + d\,, \tag{3.32}$$

using the notation

$$a = \frac{C}{A-v}\,,\quad b = \frac{CD}{v(A-v)}\,,\quad c = \frac{B(A-v) + CC_1}{v(A-v)}\,,\quad d = -\frac{D}{v}\,.$$

It is worth noting that $b = -ad$. This allows us to find a partial solution to (3.32) in the form $n_0 = ds - c/a$ (see [369], p. 13, No. 17), and further, its general solution which is given by (see [369], p. 10)

$$n(s) = n_0(s) + \Phi(s)\left[C_2 - a\int \Phi(s)\mathrm{d}s\right]^{-1}\,. \tag{3.33}$$

In our case, the function $\Phi(s)$ is easy to find as

$$\Phi(s) = \exp\left(-\frac{b}{2}s^2 - cs\right)\,, \tag{3.34}$$

and C_2 is a constant. By introducing (3.34) into (3.33), we find a general solution for (3.32) in the case $b = -ad$:

$$n(s) = ds - \frac{c}{a} + \frac{\exp\left(-\frac{b}{2}s^2 - cs\right)}{C_2 - a\sqrt{\frac{\pi}{2b}}\exp\left(\frac{c^2}{2b}\right)\mathrm{erf}\left(s\sqrt{\frac{b}{2}} + \frac{c}{\sqrt{2b}}\right)}\,. \tag{3.35}$$

Coming back to our initial notation, we obtain the solution for the system (3.29)–(3.30) in the following closed form [368]:

$$\rho(s) = -\frac{B}{C} + \frac{v}{A-v}\frac{P(s)}{C_2 - Q(s)}\,, \tag{3.36}$$

$$n(s) = -\frac{(A-v)B + CC_1}{vC} - \frac{D}{v}s + \frac{P(s)}{C_2 - Q(s)}\,, \tag{3.37}$$

where the following functions

$$P(s) = \exp \frac{-CDs^2 - 2[(A-v)B + CC_1]s}{2v(A-v)}\,, \tag{3.38}$$

$$Q(s) = \sqrt{\frac{\pi v C}{2(A-v)D}} \exp \frac{[(A-v)B + CC_1]^2}{2v(A-v)CD} \operatorname{erf} \frac{CDs + (A-v)B + CC_1}{\sqrt{2v(A-v)CD}} \tag{3.39}$$

have been introduced.

To analyse the solution in detail, one has to estimate the main parameters A, B, C and D, which must depend strongly on experimental conditions. At the moment, there are no reliable data for making such estimates, so our theoretical results must await verification in future experiments.

One can draw the following conclusions from the results of Sect. 3.1:

- Homogeneous and inhomogeneous plastic flow in NCMs can be effectively described by means of the concept of cooperative grain boundary sliding and the model of cellular dislocations.
- Elastic fields of cellular dislocations can be modeled with the help of dislocation solutions obtained within the theory of gradient elasticity which allows one to avoid classical singularities in displacements, strains, stresses and energies at the dislocation lines, and to study short-range elastic interactions between dislocations when they coalesce or split.
- The theoretical model [351] describing nucleation and motion of cellular dislocations under the action of low- or high-intensity accommodation processes shows a possible correlation between these accommodation processes and the realization of respectively homogeneous or inhomogeneous plasticity regimes.
- The efficiency of the proposed approach has been demonstrated for the concrete example of TiB_2 and TiN nanostructured films, where activation of accommodation processes of different intensity is related to the difference in anisotropy and dislocation slip systems between the hexagonal (TiB_2) and cubic (TiN) crystalline lattices.
- The spatiotemporal evolution of an ensemble of cellular dislocations can be described by a system of two coupled differential chemical-reaction-kinetics equations with respect to the densities of mobile and immobile cellular dislocations, which has an exact solution in a mobile coordinate system.
- A detailed analysis of the solution requires further experimental studies of cellular dislocation behavior.

3.2 Concept of Grain Boundary Dislocations

NCMs are often extremely hard and brittle, but several examples of substantial ductility under mechanical load have been reported. Some nanocrystalline alloys even exhibit superplasticity at relatively high strain rates and low temperatures, which is characterized by very high flow stresses and strengthening at the first extensive stage of deformation [336–338, 370–375]. In other words, such NCMs are simultaneously superplastic and superstrong. Very high values of the flow stress and strengthening at the extensive stage of deformation are specific features of superplastic NCMs, distinguishing their deformation behavior from that of conventional microcrystalline materials exhibiting superplasticity. These features are a subject of growing fundamental interest [336–338, 370–375], motivated by a range of new commercial applications using the superplasticity of NCMs.

After the first extensive stage of superplastic deformation characterized by strengthening, a second stage occurs which is characterized by softening and plastic flow localization, and ends by failure [337, 338, 371–375]. Very large values of the flow stress as well as the strengthening and softening effects in NCMs under superplastic deformation can be naturally attributed to the action of grain boundaries whose volume fraction is extremely high in such materials (see Sect. 1.2). In particular, grain boundary sliding is considered to be the dominant mode of superplasticity in NCMs [338, 374, 375], in which case the unusual strengthening and very high values of the flow stress should be related to the specific peculiarities of grain boundary sliding in NCMs. The main aim of this section is to discuss the theoretical model describing the strengthening [376, 377] and softening [377] mechanisms in NCMs under superplastic deformation as those related to transformations of grain boundary dislocations – elementary carriers of grain boundary sliding – at triple junctions of grain boundaries.

3.2.1 Transformations of Grain Boundary Dislocations at Triple Junctions

Following [376, 377], consider the evolution of grain boundary defects in an NCM under mechanical load. Grain boundaries with excess density of grain boundary dislocations often exist in as-fabricated NCMs [320, 371]. When a mechanical load is applied to the specimen, the mobile grain boundary dislocations (with Burgers vectors parallel to the grain boundary planes) begin to move, thus causing grain boundary sliding. They are stopped at triple junctions of grain boundaries that represent effective obstacles for dislocation movement (Fig. 3.11a). In general, the grain boundary dislocations stopped near a triple junction are capable of overcoming the junction obstacle and come into a dislocation reaction when the shear stress reaches some critical value. In NCMs with their high-density ensembles of triple junctions, the critical shear stress needed for grain boundary dislocations to overcome triple junctions specifies the contribution of grain boundary sliding to the yield stress. Following [338, 374, 375], we assume that grain boundary sliding is the dominant mode of superplasticity, in which case the critical shear stress causes the flow stress of superplastic deformation in NCMs. Therefore, the dependence of the critical stress on parameters of defect structures is responsible for the basic peculiarities of the superplastic flow here.

The authors [376, 377] have calculated these dependencies using the first approximation model that describes transformations of grain boundary dislocations at triple junctions. Let us consider a model configuration of grain boundary dislocations in an NCM, which is formed near a triple junction under the action of shear stress τ (Fig. 3.11a) [376,377]. The configuration consists of two grain boundary dislocations with Burgers vectors $\boldsymbol{b}_1$ and $-\boldsymbol{b}_2$ ($\boldsymbol{b}_1$- and $-\boldsymbol{b}_2$-dislocations) which are parallel to the corresponding grain boundary planes adjacent to the triple junction. The grain boundary dislocations are stopped by the triple junction when $\tau < \tau_1^{\mathrm{c}}$. In order to estimate the yield stress for grain boundary sliding, one can analyze the energetic characteristics of transformations of the grain boundary dislocations occurring at the triple junction. In doing so, it should be noted that an elementary act of grain boundary sliding is a transfer of the $-\boldsymbol{b}_2$-dislocation across the triple junction (Figs. 3.11a and b). This transfer over a short distance l_2 occurs at the critical shear stress τ_1^{c} at which the transfer becomes energetically favorable. The transfer is accompanied by a dislocation reaction which involves the $\boldsymbol{b}_1$- and $-\boldsymbol{b}_2$-dislocations and results in both the formation of a sessile grain boundary dislocation with Burgers vector $\boldsymbol{b} = \boldsymbol{b}_1 - \boldsymbol{b}_2$ ($\boldsymbol{b}$-dislocation) and a displacement of the triple junction by the vector $\boldsymbol{b}_2$ (Fig. 3.11b). In the framework of the model [376,377], the process discussed is an elementary act of superplastic deformation of a nanocrystalline specimen under mechanical load.

After the dislocation reaction has occurred at the triple junction (see Figs. 3.11a and b), the two new grain boundary dislocations are generated, say, as those resulting from dissociation of lattice dislocations trapped by grain boundaries. The new grain boundary dislocations move under shear stress action towards the triple junction (Fig. 3.11c). (For simplicity, it is first assumed that all moving grain boundary dislocations at the boundaries AO and OB are characterized by Burgers vectors $\boldsymbol{b}_1$ and $-\boldsymbol{b}_2$, respectively.) These moving $\boldsymbol{b}_1$- and $-\boldsymbol{b}_2$-dislocations interact elastically with the sessile $\boldsymbol{b}$-dislocation, which hampers the movement of the new dislocations. At the stress level $\tau_2^{\mathrm{c}} > \tau_1^{\mathrm{c}}$, a new elementary act of grain boundary sliding occurs at the triple junction. With the Burgers vector conservation law, the elementary act results in the formation of a new sessile $\boldsymbol{b}$-dislocation and a new transfer of the triple junction by $\boldsymbol{b}_2$ (Fig. 3.11d). The new and pre-existing sessile $\boldsymbol{b}$-dislocations converge, resulting in the formation of a new sessile dislocation with Burgers vector $2\boldsymbol{b} = 2(\boldsymbol{b}_1 - \boldsymbol{b}_2)$.

The process under consideration occurs repeatedly and is accompanied by an increase in the Burgers vector of the sessile dislocation by the vector $\boldsymbol{b} = \boldsymbol{b}_1 - \boldsymbol{b}_2$ at each step. Since the sessile dislocation hampers moving GB dislocations, the critical shear stress increases with evolution of the defect structure. We think that this phenomenon is responsible for the experimen-

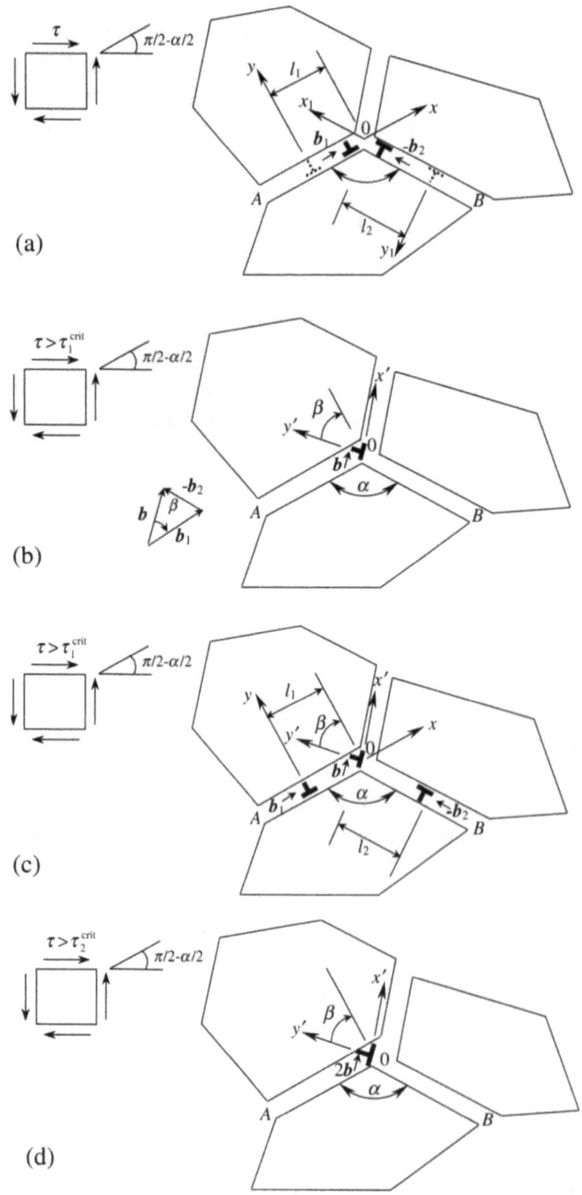

Fig. 3.11. Transformations of grain boundary dislocations near a triple junction. (**a**) The initial (0 th) state of the defect configuration. Two gliding grain boundary dislocations move towards triple junction O. (**b**) A sessile dislocation with Burgers vector $\boldsymbol{b}$ is formed. The triple junction is transferred by vector $\boldsymbol{b}_2$ from its initial position. (**c**) Generation of two new gliding grain boundary dislocations with Burgers vectors $\boldsymbol{b}_1$ and $-\boldsymbol{b}_2$, which move towards the triple junction. (**d**) A sessile dislocation with Burgers vector $\mathbf{2}b$ is formed. The triple junction is transferred by vector $\mathbf{2}b_2$ from its initial position

tally detected strengthening [337, 338, 371–375] in NCMs under superplastic deformation (occurring mostly through grain boundary sliding).

In addition to the formation of sessile grain boundary dislocations with growing Burgers vectors, the passage of grain boundary dislocations through the triple junction gives rise to a transfer of the triple junction itself (Figs. 3.11 and 3.12). More precisely, in the framework of the model [377], n elementary acts of grain boundary sliding at the triple junction cause the junction to transfer by $n\boldsymbol{b}_2$ (Fig. 3.12). In its turn the transfer is capable of giving rise to local grain boundary migration driven by a decrease in the summed length of grain boundaries and a corresponding decrease in their energy. This phenomenon and its softening effect in NCMs under superplastic deformation is discussed in detail below.

3.2.1.1 Energetics of the First Transformation Event

Let us consider the energy characteristics of the first event of transformation of grain boundary dislocations at a triple junction (Figs. 3.11a and b), which occurs under the action of a shear stress τ. The first transformation is characterized by the difference ΔW_1 between the energies of the final (W_1) and initial (W_0) states of the defect configuration, shown in Figs. 3.11b and a, respectively. It is energetically favorable (unfavorable), if $\Delta W_1 < 0$ ($\Delta W_1 > 0$). The equation $\Delta W_1 = 0$ gives a set of critical values of parameters for the defect configuration, at which its transformation becomes energetically favorable.

The initial defect configuration represents two gliding grain boundary dislocations stopped near a triple junction (Fig. 3.11a). Its energy W_0 comprises three terms [377]:

$$W_0 = E_{\mathrm{s}}^{b_1} + E_{\mathrm{s}}^{b_2} + E_0^{b_1-b_2} \,, \tag{3.40}$$

where $E_{\mathrm{s}}^{b_1}$ and $E_{\mathrm{s}}^{b_2}$ are the self-energies of the $\boldsymbol{b}_1$- and $-\boldsymbol{b}_2$-dislocations, respectively, and $E_0^{b_1-b_2}$ is the energy of elastic interaction between them.

The self-energies are given by the standard formula [349]

$$E_{\mathrm{s}}^{b_i} = \frac{Db_i^2}{2}\left(\ln\frac{R}{r_{\mathrm{c}_i}} + 1\right) \,, \tag{3.41}$$

where $D = G/[2\pi(1-\nu)]$, G denotes the shear modulus, ν the Poisson ratio, r_{c_i} the cutoff radius of the stress field of the dislocations with Burgers vectors $\boldsymbol{b}_1$ and $-\boldsymbol{b}_2$, for $i = 1$ and 2, respectively, and R denotes the screening length of the dislocation stress field.

The energy of elastic interaction between two defects can be calculated as the work spent to transfer one defect from a free surface of a solid to its current position in the stress field created by another defect [378]. In this approach, for our case with two grain boundary dislocations (Fig. 3.11a), their interaction energy $E_0^{b_1-b_2}$ can be calculated using the formula

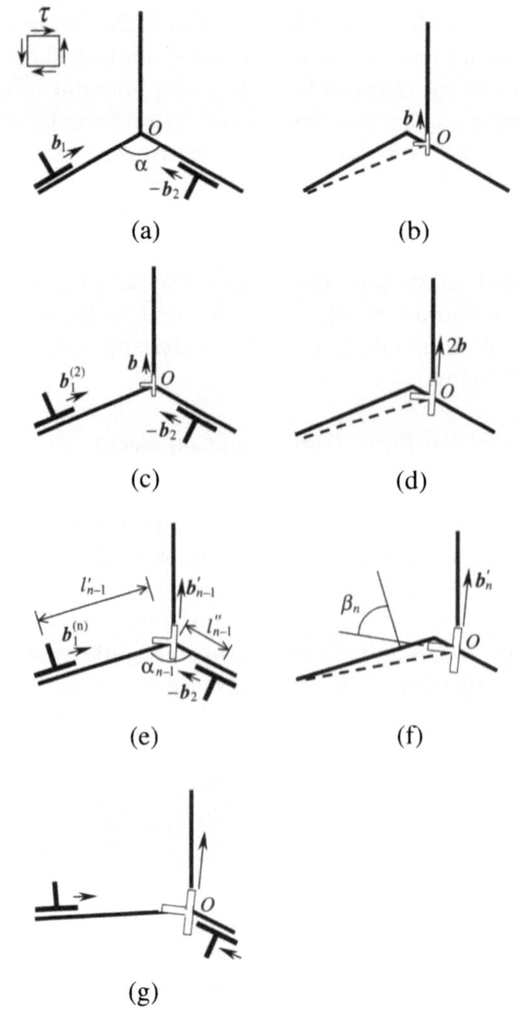

Fig. 3.12. Numerous transfers of grain boundary dislocations across the triple junction and accompanying local migration of grain boundaries lead to an increase in the angle α

$$E_0^{b_1-b_2} = -b_1 \int_{-R}^{0} \tau_{xy}^{b_2}(x, y = 0)\mathrm{d}x \,, \tag{3.42}$$

where $\tau_{xy}^{b_2}$ is the shear stress induced by the $-\boldsymbol{b}_2$-dislocation and acting on the $\boldsymbol{b}_1$-dislocation. This stress is written as

$$\tau_{xy}^{b_2}(x, y) = \frac{1}{2}\left(\sigma_{x_1x_1}^{b_2} - \sigma_{y_1y_1}^{b_2}\right)\sin(2\alpha) + \sigma_{x_1y_1}^{b_2}\cos(2\alpha) \,. \tag{3.43}$$

where the dislocation stress components $\sigma^{b_2}_{x_1x_1}$, $\sigma^{b_2}_{y_1y_1}$ and $\sigma^{b_2}_{x_1y_1}$ are written in the coordinate system (x_1, y_1) associated with the Burgers vector $-\boldsymbol{b}_2$ and rotated by α relative to the coordinate system (x, y) associated with the Burgers vector $\boldsymbol{b}_1$ (Fig. 3.11a). These components are given (in units of Db_2) by the well-known formulas [349]

$$\sigma^{b_2}_{x_1x_1} = -y_1\frac{y_1^2+3x_1^2}{r_1^4}\,, \quad \sigma^{b_2}_{y_1y_1} = y_1\frac{x_1^2-y_1^2}{r_1^4}\,, \quad \sigma^{b_2}_{x_1y_1} = x_1\frac{x_1^2-y_1^2}{r_1^4}\,, \tag{3.44}$$

with $r_1^2 = x_1^2 + y_1^2$. In (3.44), the coordinates (x_1, y_1) bear the following relationships with the coordinates (x, y):

$$x_1 = -(x-x_0)\cos\alpha + (y+y_0)\sin\alpha\,, \quad y_1 = (x-x_0)\sin\alpha + (y+y_0)\cos\alpha\,, \tag{3.45}$$

where $x_0 = l_1 - l_2\cos\alpha$ and $y_0 = l_2\sin\alpha$. Here l_i is the distance from the source of the grain boundary dislocation to the triple junction under consideration.

Using (3.44) and (3.45) and substituting (3.43) into (3.42) yields [377]

$$E_0^{b_1-b_2} = Db_1b_2\left\{\frac{1}{2}\cos\alpha\ln\left[1+\frac{R^2+2R(l_1-l_2\cos\alpha)}{l_1^2+l_2^2-2l_1l_2\cos\alpha}\right]\right.$$
$$\left.-\frac{l_1l_2\sin^2\alpha}{l_1^2+l_2^2-2l_1l_2\cos\alpha}+\frac{2l_2(R+l_1)\sin^2\alpha}{R^2+2R(l_1-l_2\cos\alpha)+l_1^2+l_2^2-2l_1l_2\cos\alpha}\right\}. \tag{3.46}$$

The energy of the defect configuration in its final state (Fig. 3.11b) comprises three terms [377],

$$W_1 = E_{\mathrm{s}}^b + W_{\mathrm{b}} + A\,, \tag{3.47}$$

where E_{s}^b denotes the self-energy of the sessile $\boldsymbol{b}$-dislocation, W_{b} the energetic barrier for the movement of the $-\boldsymbol{b}_2$-dislocation across the triple junction, and A the work done by the shear stress τ to transfer the $\boldsymbol{b}_1$-dislocation over the distance l_1 and the $-\boldsymbol{b}_2$-dislocation over the distance l_2. The term E_{s}^b is given by (3.41), replacing b_i and r_{c_i} by b and r_{c}, respectively. The energetic barrier for grain boundary dislocation movement across the triple junction is assumed to be $W_{\mathrm{b}} = Gb_2^2\kappa$, where κ is the fitting parameter, of the order of unity. The work of the shear stress τ is written in the standard way as $A = \tau(b_1l_1 + b_2l_2)\cos\alpha$.

From Figs. 3.11a and b it follows that the Burgers vector magnitude b characterizing the sessile dislocation bears the following relationships with the magnitudes b_1 and b_2 of the Burgers vectors of the moving grain boundary dislocations:

$$b_1 = b\cos\beta - b_2\cos\alpha \quad \text{and} \quad b_2\sin\alpha = b\sin\beta\,.$$

Here β is the angle by which the coordinate system (x', y') associated with the sessile dislocation is rotated relative to the coordinate system (x, y) associated with the moving $\boldsymbol{b}_1$-dislocation. These relationships can be rewritten in the form

$$b = b_2 \frac{\sin\alpha}{\sin\beta} \quad \text{and} \quad \beta = \operatorname{arcctg}\left(\frac{b_1}{b_2}\operatorname{cosec}\alpha + \operatorname{ctg}\alpha\right) .$$

Using (3.40), (3.41), (3.46) and (3.47), one can find an expression for the characteristic energy difference $\Delta W_1 = W_1 - W_0$ [377]:

$$\Delta W_1 = Gb_2^2\kappa + \tau(b_1 l_1 + b_2 l_2)\cos\alpha + \frac{D}{2}\left\{(b^2 - b_1^2 - b_2^2)\left(\ln\frac{R}{r_c} + 1\right)\right.$$
$$-b_1 b_2\left[\cos\alpha \ln\left(1 + \frac{R^2 + 2R(l_1 - l_2\cos\alpha)}{l_1^2 + l_2^2 - 2l_1 l_2\cos\alpha}\right) + \frac{l_1 l_2 \sin^2\alpha}{l_1^2 + l_2^2 - 2l_1 l_2\cos\alpha}\right.$$
$$\left.\left. - \frac{4l_2(R + l_1)\sin^2\alpha}{R^2 + 2R(l_1 - l_2\cos\alpha) + l_1^2 + l_2^2 - 2l_1 l_2\cos\alpha}\right]\right\} , \qquad (3.48)$$

where the approximation $r_{c_i} \approx r_c$ has been used.

The energy difference ΔW_1 has been analysed numerically with regard to its dependence on the applied shear stress τ for different values of the angle α and the following values of the model parameters: $l_1 = l_2 = 100b$, $R = 10^7 b$, and $\kappa = 1$ (Fig. 3.13) [377]. As can be seen from the plots, ΔW_1 decreases with rising τ and also with rising α. It is positive (the transformation is energetically unfavorable) for relatively small values of τ and α, and negative (the transformation is favorable) for large values of these quantities. When $\alpha \leq 100°$, one has to apply extremely high ($> G/50$) shear stress to make the transformation energeticaly favorable. When $\alpha \geq 140°$, the first transformation event is favorable, even without external stress. (Of course, provided that the grain boundary dislocations already exist near the triple junction.) In the intermediate range of α (here, for $\alpha = 120°$), it is enough

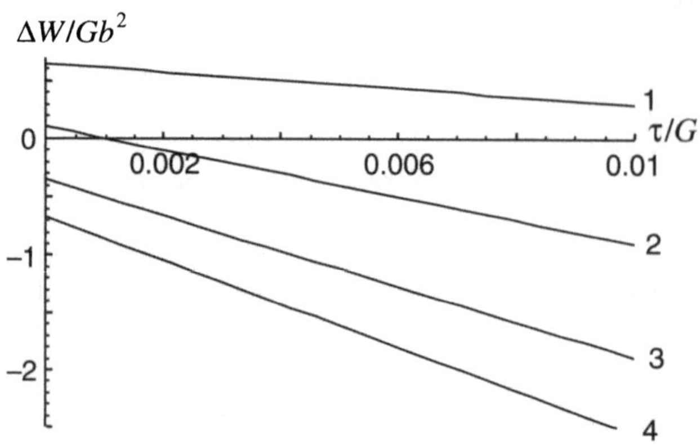

Fig. 3.13. Dependence of the energy difference ΔW_1 on the external shear stress τ for different values of the angle $\alpha = 100°$ (1), $120°$ (2), $140°$ (3), and $160°$ (4)

to apply some reasonable critical shear stress (here $\tau_1^c \approx G/330$) to activate the first transformation event. It is also evident that the level of τ_1^c increases with decreasing α.

3.2.1.2 Energetics of the n th Transformation Event

The authors [377] have thus shown that the transformations of grain boundary dislocations at triple junctions are energetically favorable in a wide range of the model parameters. Let us consider how the situation develops further, with the increasing Burgers vector of the sessile dislocation during subsequent transfers of grain boundary dislocations through the triple junction and the corresponding transfer of the triple junction itself. It is worth noting now that, in the process of the triple junction transfer, the adjacent grain boundaries change their orientation and hence the geometry of the defect configuration changes after every elementary act of grain boundary sliding. This means that the Burgers vectors $\boldsymbol{b}_1$ and $-\boldsymbol{b}_2$ change their orientation together with the corresponding grain boundaries, the angle α increases, the distance l_1 increases, and the distance l_2 decreases. All these changes greatly influence the energetics of dislocation transformations at the transferred triple junction. These changes are not significant for the first transformation events because the characteristic magnitude of the Burgers vector $-\boldsymbol{b}_2$ is of the order of 0.1 nm [365], while the thickness of the grain boundaries and their triple junctions is about 1 nm in NCMs (see Sect. 1.2). Therefore, at least the first 10 transformation events have practically no influence on the system geometry. However, under superplastic deformation, one can expect hundreds of such events and this will make the system geometry vary significantly.

Now let us examine the energy characteristics of the transformation of the grain boundary dislocations at a triple junction containing a sessile dislocation with the Burgers vector $\boldsymbol{b}'_{n-1}$ formed after $(n-1)$ transformation events. In fact, it is the n th act of superplastic deformation occurring at the triple junction and resulting in the formation of the sessile dislocation with the Burgers vector $\boldsymbol{b}'_n$. The corresponding angles between the grain boundaries are now α_{n-1} and α_n. The corresponding positions of grain boundary dislocation sources with respect to the triple junction are now l'_{n-1} (instead of l_1), l''_{n-1} (instead of l_2) and l'_n, l''_n, respectively. The relationships between the 'old' (initial) and 'new' (current) parameters can be obtained from Fig. 3.14 and the obvious equations

$$\begin{aligned} L^2 &= 2d^2(1-\cos\alpha) , \\ L^2 &= d'^2 + (d-nb_2)^2 - 2d'(d-nb_2)\cos\alpha_n , \\ d'^2 &= d^2 + n^2b_2^2 - 2ndb_2\cos\alpha , \end{aligned} \tag{3.49}$$

where L is the distance between the neighbouring triple junctions, d the initial length of both the grain boundaries under discussion, and d' the current length of the boundary where the $\boldsymbol{b}_1$-dislocation glides. As a result, one can derive the formulas [377]

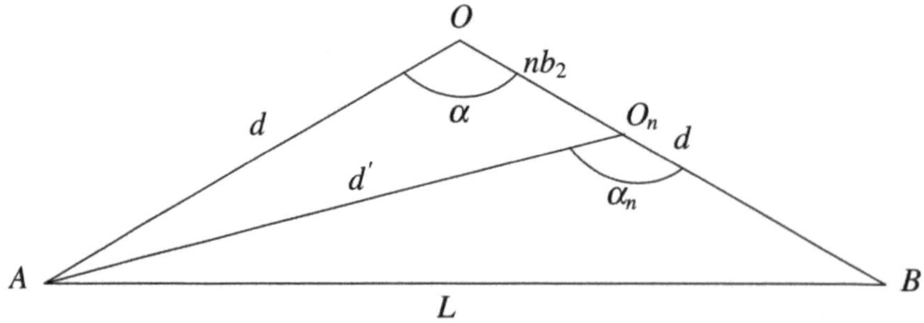

Fig. 3.14. Geometry of triple junction transfer

$$l'_n = \frac{l_1}{d}\sqrt{d^2 + n^2b_2^2 - 2ndb_2\cos\alpha} ,$$
$$l''_n = l_2 - nb_2 , \qquad (3.50)$$
$$\alpha_n = \arccos\frac{n^2b_2^2 - nb_2d + d(d - nb_2)\cos\alpha}{(d - nb_2)\sqrt{d^2 + n^2b_2^2 - 2ndb_2\cos\alpha}} .$$

The magnitude and orientation of the Burgers vector $\boldsymbol{b}'_n$ of the sessile dislocation is then determined by

$$b'_n = b_2\sum_{i=1}^{n}\sin\alpha_i \operatorname{cosec}\beta_i , \qquad (3.51)$$

$$\beta_n = \operatorname{arcctg}\left(\frac{b_1}{b_2}\operatorname{cosec}\alpha_n + \operatorname{ctg}\alpha_n\right) . \qquad (3.52)$$

The transformation in question is characterized by the difference ΔW_n in the energy between the n th (W_n) and $(n-1)$ th (W_{n-1}) states of the defect system. The transformation is energetically favorable if $\Delta W_n < 0$. The set of critical parameters, including the critical shear stress τ_n^{c} for grain boundary sliding across the triple junction with the sessile $\boldsymbol{b}'_n$-dislocation, is derived from the equation $\Delta W_n = 0$.

The energy W_{n-1} (per unit length of a grain boundary dislocation) comprises six terms [377]:

$$W_{n-1} = E_{\mathrm{s}}^{b_1} + E_{\mathrm{s}}^{b_2} + E_{\mathrm{s}}^{b'_{n-1}} + E_{n-1}^{b_1-b_2} + E_{n-1}^{b_1-b'_{n-1}} + E_{n-1}^{b_2-b'_{n-1}} . \qquad (3.53)$$

Here $E_{\mathrm{s}}^{b_1}$, $E_{\mathrm{s}}^{b_2}$ and $E_{\mathrm{s}}^{b'_{n-1}}$ are the self-energies of the $\boldsymbol{b}_1$-, $-\boldsymbol{b}_2$- and $\boldsymbol{b}'_{n-1}$-dislocations, respectively, $E_{n-1}^{b_1-b_2}$ is the energy of elastic interaction between the mobile $\boldsymbol{b}_1$- and $-\boldsymbol{b}_2$-dislocations, and $E_{n-1}^{b_1-b'_{n-1}}$ ($E_{n-1}^{b_2-b'_{n-1}}$) the energy of elastic interaction between the mobile $\boldsymbol{b}_1$- ($-\boldsymbol{b}_2$-) and sessile $\boldsymbol{b}'_{n-1}$-dislocations.

The self-energies $E_{\mathrm{s}}^{b_1}$ and $E_{\mathrm{s}}^{b_2}$ are given by (3.41), and $E_{\mathrm{s}}^{b'_{n-1}}$ by (3.41), replacing b_i and r_{c_i} by b'_{n-1} and $r'_{\mathrm{c}n-1}$, respectively.

The interaction energy $E_{n-1}^{b_1-b_2}$ is obtained from (3.46) by using the substitutions $\alpha \rightarrow \alpha_{n-1}$, $l_1 \rightarrow l'_{n-1}$, and $l_2 \rightarrow l''_{n-1}$. The interaction energies $E_{n-1}^{b_1-b'_{n-1}}$ and $E_{n-1}^{b_2-b'_{n-1}}$ are calculated in a way similar to that used in deriving (3.46). This gives [377]

$$E_{n-1}^{b_1-b'_{n-1}} = Db_1b'_{n-1}\cos\beta_{n-1}\ln\left(\frac{R}{l'_{n-1}}+1\right) , \tag{3.54}$$

$$E_{n-1}^{b_2-b'_{n-1}} = Db_2b'_{n-1}\cos(\alpha_{n-1}-\beta_{n-1})\ln\left(\frac{R}{l''_{n-1}}+1\right) . \tag{3.55}$$

The energy W_n of the defect configuration formed after n transfers of the grain boundary dislocations across the triple junction is given by (3.47) with the replacements $b \rightarrow b'_n$, and $r_c \rightarrow r'_{c_n}$ in its first term, and $\alpha \rightarrow \alpha_{n-1}$, $l_1 \rightarrow l'_{n-1}$, and $l_2 \rightarrow l''_{n-1}$ in its third term.

The final result for the energy difference $\Delta W_n = W_n - W_{n-1}$ then reads [377]

$$\begin{aligned}
\Delta W_n &= Gb_2^2\kappa + \tau(b_1l'_{n-1} + b_2l''_{n-1})\cos\alpha_{n-1} \\
&+\frac{D}{2}\Bigg\{b_n'^2\left(\ln\frac{R}{r'_{c_n}}+1\right) - b_{n-1}'^2\left(\ln\frac{R}{r'_{c_{n-1}}}+1\right) - (b_1^2+b_2^2)\left(\ln\frac{R}{r_c}+1\right) \\
&-b_1b_2\Bigg[\cos\alpha_{n-1}\ln\left(1+\frac{R^2+2R(l'_{n-1}-l''_{n-1}\cos\alpha_{n-1})}{l_{n-1}'^2+l_{n-1}''^2-2l'_{n-1}l''_{n-1}\cos\alpha_{n-1}}\right) \\
&+\frac{l'_{n-1}l''_{n-1}\sin^2\alpha_{n-1}}{l_{n-1}'^2+l_{n-1}''^2-2l'_{n-1}l''_{n-1}\cos\alpha_{n-1}} \\
&-\frac{4l''_{n-1}(R+l'_{n-1})\sin^2\alpha_{n-1}}{R^2+2R(l'_{n-1}-l''_{n-1}\cos\alpha_{n-1})+l_{n-1}'^2+l_{n-1}''^2-2l'_{n-1}l''_{n-1}\cos\alpha_{n-1}}\Bigg] \\
&-2b'_{n-1}\left[b_1\cos\beta_{n-1}\ln\left(\frac{R}{l'_{n-1}}+1\right)+b_2\cos(\alpha_{n-1}-\beta_{n-1})\ln\left(\frac{R}{l''_{n-1}}+1\right)\right]\Bigg\} .
\end{aligned} \tag{3.56}$$

3.2.2 Strengthening and Softening Under Superplastic Deformation

Let us calculate the critical shear stress τ_n^c needed for the $-\boldsymbol{b}_2$-dislocation to move across the triple junction containing the sessile $\boldsymbol{b}'_n$-dislocation. This stress can be found from the condition that $\Delta W_n = 0$. In doing so, we have [377]

$$\tau_n^c = \frac{E_s^{b'_n} - E_s^{b'_{n-1}} - E_s^{b_1} - E_s^{b_2} - E_{n-1}^{b_1-b_2} - E_{n-1}^{b_1-b'_{n-1}} - E_{n-1}^{b_2-b'_{n-1}} + W_b}{(b_1l'_{n-1}+b_2l''_{n-1})\cos\alpha_{n-1}} . \tag{3.57}$$

Any real nanocrystalline sample is characterized by a distribution of geometric parameters (in particular, α) describing its triple junctions. We have averaged the shear stress τ_n^{c} over the angle α that is specified by some distribution in a mechanically loaded nanocrystalline sample. The angle distribution was assumed to be normal, viz.,

$$f(\alpha) = \frac{1}{s_\alpha\sqrt{2\pi}} \exp \frac{-(\alpha-\xi)^2}{2s_\alpha^2} , \tag{3.58}$$

where ξ and s_α are the distribution center and standard deviations of α, respectively.

As is shown in Sect. 3.2.1.1, the transformations of grain boundary dislocations at triple junctions are more energetically favorable for triple junctions characterized by larger angles α, when $\alpha \geq 140°$. Therefore, it seems reasonable to take the range of α from 140° to 180°. In this characteristic range, its mean value is given by

$$m_\alpha = \int\limits_{7\pi/9}^{\pi} \alpha f(\alpha) \mathrm{d}\alpha . \tag{3.59}$$

Hence, the critical stress is determined, not from the average angle, but from the whole distribution:

$$\bar{\tau}_n^{\mathrm{c}} = \frac{1}{m_\alpha} \int\limits_{7\pi/9}^{\pi} \alpha f(\alpha) \tau_n^{\mathrm{c}}(\alpha) \mathrm{d}\alpha . \tag{3.60}$$

Using (3.57)–(3.60), we have numerically calculated the dependence of $\bar{\tau}_n^{\mathrm{c}}$ on the number of transformation events n, for a normally distributed angle α that characterizes the geometry of the triple junction, with $\xi = 160°$ and $s_\alpha = 10°$. The following values for the system parameters were used in the calculations: $G = 70$ GPa (for pure nickel), $\nu = 0.3$, $b_1 = b_2 = r_{\mathrm{c}} = 0.1$ nm, $r'_{\mathrm{c}_n} = b'_n$, $R = 10^7$ nm, and $\kappa = 1.75$. (This value of κ corresponds to the condition $\bar{\tau}_n^{\mathrm{c}} \geq 0$ for the given values of the other parameters.) The distances d and $l_1 = l_2 = l$ were varied in the calculations.

The plots $\bar{\tau}_n^{\mathrm{c}}(n)$ are presented in Fig. 3.15, showing that the mean critical stress first increases with rising n, achieves its maximum value, and subsequently decreases to zero. The number n of elementary acts of superplastic deformation occurring at a triple junction serves as a characteristic of the degree of local plastic strain at this triple junction. In this context, the dependences $\bar{\tau}_n^{\mathrm{c}}(n)$ have the same character as the 'flow stress–strain' dependences in real NCMs under superplastic deformation, thus demonstrating the strengthening and softening detected experimentally in [336–338, 370–375].

Within the model [377], the strengthening stage is explained by a strong elastic interaction between mobile grain boundary dislocations gliding towards the triple junction and the sessile grain boundary dislocation with increasing Burgers vector which is the product of the dislocation transfer across

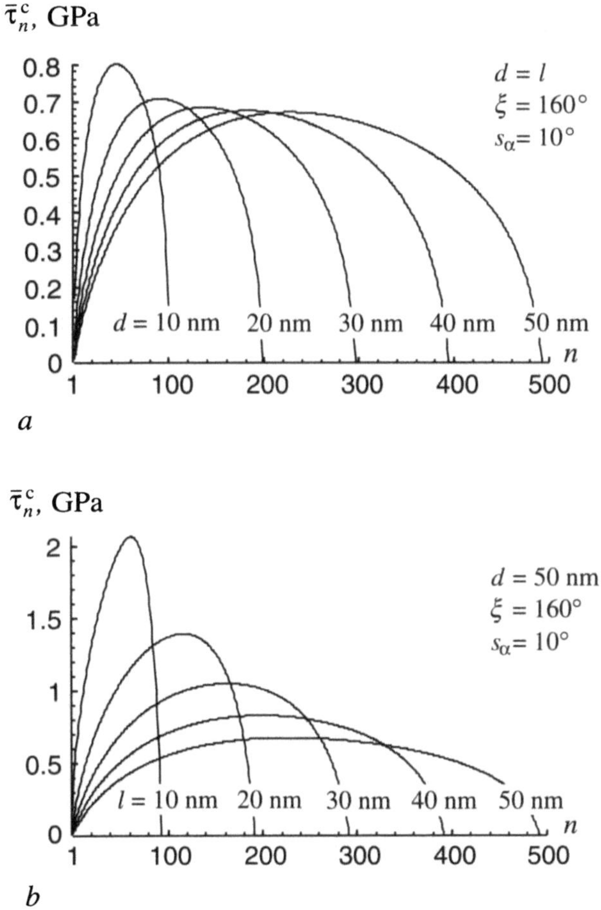

Fig. 3.15. Dependence of the mean critical shear stress $\bar{\tau}_n^c$ on the number of transformation events n (**a**) for different values of the distance $l = d$, and (**b**) for different values of the distance $l \leq d$, where $d = 50$ nm. The values of l are shown in nanometers on the curves

the triple junction. The softening stage arises for geometric reasons when the triple junction migrates and the angle α increases. This changes the orientation of the Burgers vectors of the gliding grain boundary dislocations. They begin to interact elastically more strongly with each other (attracting each other) and less strongly with the sessile dislocation. Furthermore, the action of the applied shear stress on them becomes more intensive. In these circumstances, triple junctions stop being geometric obstacles for the movement of new grain boundary dislocations, and this movement is therefore enhanced along the shear surface. As a result, a smaller stress is needed to make the grain boundary dislocation pass through the triple junction.

The enhancement of grain boundary sliding due to local migration of grain boundaries has been discussed in [307, 379], which focus on a qualitative description of this phenomenon. The discussed representations of the softening mechanism related to local migration of grain boundaries are supported by experimental observation [380] of superplastic flow localization in NCMs. According to electron microscopy experiments [380], shear bands where superplastic flow is localized in NCMs contain brick-like grains with grain boundaries parallel and perpendicular to the shear direction. This is effectively and naturally interpreted as a result of grain boundary migration (Fig. 3.12) accompanying grain boundary sliding across triple junctions.

The plots $\bar{\tau}_n^{\mathrm{c}}(n)$ in Fig. 3.15 demonstrate the effect of the structural parameters d and l on the strengthening and softening regimes. In a fine-grain sample, they go faster than in a coarse-grain sample (Fig. 3.15a). The maximum value of $\bar{\tau}_n^{\mathrm{c}}$ is higher in the material with finer grains, although this dependence is not strong. In fact, the whole strengthening–softening process stops in this model when the shift nb_2 of the triple junction becomes equal to l_2 (Fig. 3.15). In the case $d = l$, one can roughly estimate that the maximum value of $\bar{\tau}_n^{\mathrm{c}}$ is achieved at the strain $\varepsilon \approx nb_2/d \approx 0.5$, which is approximately two times smaller than the experimentally detected value [338]. At the same time, the maximum value of $\bar{\tau}_n^{\mathrm{c}}$ is about of 700 MPa for $d = 20$–50 nm (Fig. 3.15a), which is of the same order of magnitude as the maximum stress reported for Ni_3Al–Cr alloy with grain size 50 nm [338].

From the results of Sect. 3.2, we can draw the following conclusions:

- It has been revealed theoretically that superplastic deformation occurring through grain boundary sliding in NCMs is characterized by strengthening and softening due to transformations of gliding grain boundary dislocations at triple junctions of grain boundaries.
- The transformations of grain boundary dislocations at triple junctions are energetically favorable in certain ranges of parameters of the defect configuration. The corresponding flow stress is mainly caused by the critical shear stress, which is highly sensitive to the degree of plastic strain characterized by n and increasing values of the angle α.
- The strengthening stage is explained by strong elastic interaction between mobile grain boundary dislocations gliding towards triple junctions and sessile grain boundary dislocations with increasing Burgers vectors, which are the product of the dislocation transfer across the triple junctions.
- The softening stage arises for geometric reasons, when triple junctions migrate, changing their shape in such a way as to facilitate grain boundary sliding across them.
- The above model scenario of the evolution of grain boundary dislocation structure at triple junctions qualitatively describes typical stress–strain curves of NCMs under superplastic deformation. To get a good quantitative agreement between theoretical and experimental results, further development of the model is desirable.

4 Rotational Plastic Deformation in Nanocrystalline Materials

The key features of NCMs (i.e., high density of grain boundaries and triple junctions with their generic defects like grain boundary dislocations and disclinations) which mainly determine their mechanical properties, may stimulate, under special conditions, the generation and development of a rotational mode of plastic flow. This mainly concerns NCMs fabricated under highly non-equilibrium conditions like ball milling and severe plastic deformation. In the present chapter, we consider different models of rotational plastic deformation in nano- and polycrystalline materials, when this deformation is realized by conservative motion of partial disclination dipoles capturing or issuing lattice dislocations or by the formation of multipole configurations of partial grain boundary disclinations leading to rotation of grains.

4.1 Generation and Development of Misorientation Bands

During the two last decades, the concept of disclinations has been broadly applied to treat mesoscopic substructures which are characteristic for metals and alloys under large deformations (e.g., see [56, 263–266, 381] for a review). Misorientation bands (MBs) represent one of the typical elements of such substructures. They are observed as long straight strips of material with a different crystallographic orientation from neighbouring areas of the material [56, 263, 264, 381–390]. The boundaries of such MBs (i.e., the misorientation boundaries) are often illustrated as low-angle dislocation tilt boundaries, although they have a finite thickness of 0.1–0.5 μm depending on the deformation magnitude [264], and in fact consist of high-density dislocation arrangements. A review of experimental data on MBs and other different rotational structures in various high-strength materials including submicrocrystalline metals and alloys can be found in [388, 389]. On the basis of these results and recent direct atomic-level observations of dipoles of partial disclinations in mechanically milled, nanocrystalline iron [218], one can assume the possibility of MB generation in typical NCMs under large deformations.

4.1.1 Misorientation Bands in Metals Under Large Deformations

The appearance and development of MBs represents one of the ways to rotational plastic flow in heavily deformed metals. MBs were found at the beginning of the 1960s in TEM experiments [382,383] (see also [56,263,265] for a review). In the 1970s and 1980s, they were extensively studied and directly connected with the formation of specific dislocation structures which may be described through partial disclinations (see [8,263–266,381,389–391] and references therein). Indeed, the presence of two edges of the misorientation boundaries which border the MB area allows one to introduce a corresponding dipole of partial disclinations whose strength is equal to the misorientation angle characterizing the MB [391]. Figure 4.1 shows the edges of the tilt misorientation boundaries as a two-axis dipole of partial wedge disclinations of strength ω. Such a disclination dipole is geometrically related to the MB parameters through the equation [56,263] $b\rho = b/l = 2\tan(\omega/2)$ for a dislocation tilt boundary. Here b denotes the Burgers vector magnitude of the dislocations composing the boundary, ρ is their linear density along the boundary, l is the interdislocation spacing, and ω is the angle of misorientation across the boundary. In the case of small misorientation angles $\omega \ll 1$, this relationship is transformed into $\rho = 1/l = \omega/b$. Further, the width of the MB is equal to the arm $2a$ of the disclination dipole (Fig. 4.1).

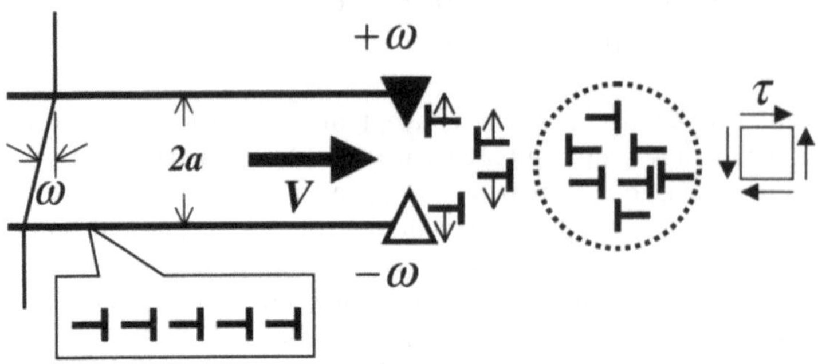

Fig. 4.1. Dislocation–disclination model of a misorientation band propagating with velocity $\boldsymbol{V}$ under the action of an external shear stress τ. The front of the misorientation band is modelled as a two-axis dipole of partial wedge disclinations of strength $\pm\omega$

In 1978, Vladimirov and Romanov proposed a dislocation–disclination model to describe the mechanism of MB propagation [391]. The main idea of the model is that the elastic stresses created by the disclination dipole (Fig. 4.1) divide a statistically arranged dislocation ensemble in front of the MB into groups of 'positive' and 'negative' dislocations. The 'positive' or 'negative' terms are caught by the positive or negative disclinations, respectively. Every event in which a dislocation dipole is captured by the disclination dipole leads to an elementary act of MB conservative motion. The mechanism [391] has been experimentally confirmed [56, 263, 264] and used to model the dislocation–disclination kinetics in metals under large deformations [56, 263–266, 381, 392–395]. However, a number of questions remain open. In particular, details of dislocation capture by disclination dipoles are still unknown. Moreover, computer simulations of elastic interactions between partial disclinations and edge dislocations [390,396] (see also Sect. 4.1.5) have shown that the simple scheme (Fig. 4.1) proposed in [391] cannot provide a complete description of MB propagation and will have to be elucidated further.

Generally speaking, the partial disclinations used to describe MBs are associated by definition with terminated misorientation boundaries in otherwise perfect crystals [56, 263]. The nature of such boundaries as well as their atomic structure may be quite different. They can be low-angle dislocation walls, high-angle grain boundaries or twin boundaries. In any case, the boundary edges may be described (both geometrically and elastically) as the lines of partial disclinations of corresponding strength. Therefore, the long-range elastic fields (far from the misorientation boundaries themselves) of displacement, strain and stress of a terminated misorientation boundary may be calculated using the well-developed mathematics of wedge disclinations within the classical [56, 263, 397, 398] or non-classical (nonlocal [147], gradient [152–154, 158] and gauge [166]) theories of elasticity.

It is well documented [56,263–266,381,384,388,389] that possible sites for MB generation in polycrystalline metals are various faults (defects) of grain boundaries including kinks, and double and triple junctions of grain boundaries. In NCMs, such grain boundary faults often contain disclinations, even in an initial as-sintered state (see Sect. 1.2). However, nowadays there is only one theoretical study [390] containing the models that allows us to describe MB generation and predict appropriate critical conditions as well as regimes of MB propagation. The well-known dislocation–disclination model of MB propagation by Vladimirov and Romanov [391] represents mainly geometrical features and needs further development.

In [390], models of initial disclination configurations at grain boundary kinks and junctions have been proposed. It has been shown that these initial configurations may serve as sources for MB generation when the applied shear stress achieves a critical value. Further development of the generated MB may be stable or unstable, depending on the level of applied shear stress. If this level is lower (higher) than another critical value, the stable (unstable) regime of MB propagation is realized.

For the case of stable propagation, an equilibrium length of the MB was introduced and studied. All these results were obtained in the framework of a quasi-equilibrium thermodynamic approach where only the necessary (not sufficient) conditions for MB generation and propagation were analyzed [390]. To obtain sufficient conditions, one should investigate the dynamics of these processes, which must be based on the dynamics of complicated dislocation structures, firstly, at the place of MB nucleation, and secondly, in front of the propagating MB. In a general three-dimensional (3D) case this problem is very complicated. However, in a 2D model case, when the lines of all dislocations and partial disclinations are assumed to be straight and parallel to each other, it can be solved by means of computer simulation within coupled dislocation–disclination dynamics [390, 396].

It is worth noting that, nowadays, computer simulation and modeling of discrete dislocation ensembles represents one of the most popular topics in theoretical materials science. Since the late 1980s, 2D and 3D calculations of the dynamics of interacting dislocations have been intensively developed (e.g., see reviews [399,400] and some recent papers on 2D [401,402] and 3D [403,404] mesoscopic simulations). However, except for [390,396], no other attempts are known to the authors which would lead to a correct simulation of dislocation–disclination ensembles. The former computer models [265, 392–395] describing the coupled evolutional kinetics of dislocations and partial disclinations have not taken into account the elastic interactions between them. In contrast, [390, 396] contain the first results of 2D computer simulations of the coupled dynamics of partial disclination dipoles and edge dislocations aimed at studying peculiarities of elastic interaction between these defects. These results have been further used to check and refine the existing theoretical (non-computer) models of MB development. In the following sections, we consider the models presented in [390, 396] in more detail.

4.1.2 Models for Initial Disclination Configurations at Grain Boundary Junctions

Consider the simple scenario [390] shown schematically in Fig. 4.2, which may be applied to initial disclination configuration formation at a grain boundary. Let some gliding dislocations with Burgers vectors $\boldsymbol{b}_1$ cross a flat grain boundary (Fig. 4.2a) and move into the neighbouring grain where the gliding

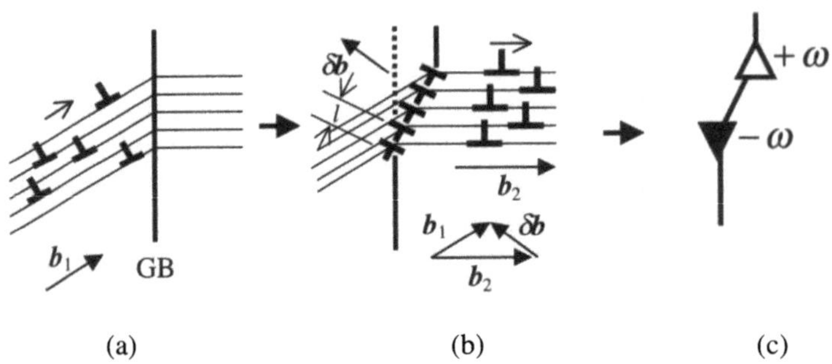

Fig. 4.2. A simple model of the formation of an initial grain boundary disclination dipole with strength ω

dislocations have Burgers vectors $\boldsymbol{b}_2$ (Fig. 4.2b). As a result, a wall of difference dislocations with Burgers vectors $\delta\boldsymbol{b} = \boldsymbol{b}_1 - \boldsymbol{b}_2$ and interdislocation spacing l appears at the crossing site, together with the grain boundary kink. On a mesoscale level, when the characteristic scale of consideration L is much larger than l, the geometry and resulting elastic fields for such a difference dislocation wall may be effectively described as those of a two-axis dipole of wedge partial disclinations with strengths $\pm\omega = \pm\delta b/l$ [56, 363, 397, 398] (Fig. 4.2c).

Such a simple scheme of grain boundary disclination dipole formation may occur in NCMs with relatively large grains which are capable of deforming by the usual glide of lattice dislocations. This may concern, for example, micro- and nanocrystalline metals and alloys fabricated by intensive plastic deformation [28, 197, 217, 218, 405]. In a real conventional polycrystalline material, the initial disclination configurations at grain boundaries result from more complicated processes, which are typical for the stage when translational deformation modes are replaced by rotational ones [56, 263, 264]. It is well established [264] that, in the vicinity of grain boundary kinks or junctions, the dislocation cells are smaller than they are when far from such sites, as is illustrated schematically in Fig. 4.3a [390]. This means that, within these areas, the density of difference grain boundary dislocations must be much higher than at straight or smooth segments of grain boundaries. As

a

b

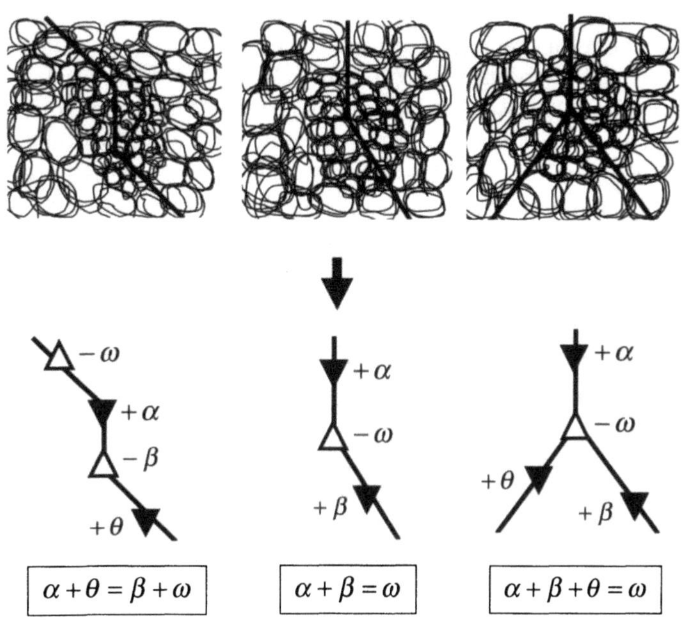

Fig. 4.3. More realistic models of the formation of initial grain boundary quadrupole-like disclination configurations

a corollary, one can assume the formation of initial quadrupole-like disclination configurations at grain boundary kinks or junctions. In Fig. 4.3b, three possible quadrupole-like disclination configurations are shown where α, β, ω and θ denote the disclination strengths. For simplicity, to catch mostly qualitative features and make rough estimates, the authors [390] considered only wedge partial disclinations. It is worth noting that the summed strength of any such disclination configuration must be equal to zero. In other words, the summed strength of positive disclinations must be equal to that of negative disclinations within a grain boundary disclination configuration (Fig. 4.3b).

4.1.3 Models for Generation of Misorientation Bands

4.1.3.1 Splitting of an Initial Grain Boundary Disclination Dipole

Consider the simplest initial grain boundary disclination configuration, that is, a grain boundary disclination dipole (Fig. 4.4a) [390]. Let this dipole be under an external shear stress τ. It is also assumed that a typical dislocation cell structure exists around the dipole. Under the action of the internal shear stress (which is caused by the grain boundary disclination dipole) and external shear stress τ, the dislocations from the cell boundaries nearest to the disclinations have to glide to them, thus forming two dislocation walls, i.e., a new disclination configuration (Fig. 4.4b) [390]. The latter may be considered as produced by a splitting of the initial grain boundary disclination dipole. This new split configuration is characterized by the split distance d, the interdislocation spacing l, the dipole arm $2a$ and the disclination strengths $\pm\alpha$ and $\pm\beta$, satisfying the relationships $\beta = b/l$ and $\omega = \alpha + \beta$, where b is the magnitude of the Burgers vector for lattice dislocations and ω is the initial strength of the dipole. In fact, the new split disclination configuration represents a model for an MB of finite length which consists of a new immobile grain boundary disclination dipole with strength α (α-dipole), a new mobile disclination dipole with strength β (β-dipole), and two misorientation boundaries of length d.

To make the transition possible from the initial grain boundary disclination dipole to the new split configuration, the total energy of the initial dipole must be larger than that of the new split configuration. Thus, to find critical conditions for such a transition, one has to calculate and compare these energies.

The total energy of the initial grain boundary disclination dipole may be calculated as the work needed to generate this dipole in its own elastic stress field [56,263]. As a result, the energy per unit length of disclinations reads

$$W_1 = D\omega^2 a^2 \left(2\ln\frac{R}{2a} + 1\right) , \qquad (4.1)$$

where $D = G/[2\pi(1-\nu)]$, G is the shear modulus, ν is the Poisson ratio, and R is a characteristic screening parameter for the disclination long-range elastic fields (e.g., the size of a sample).

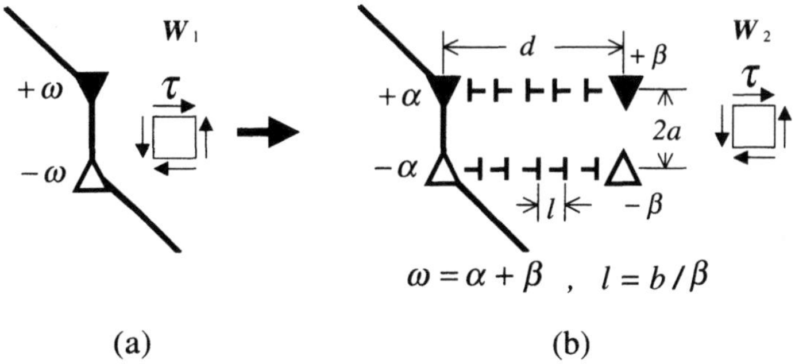

Fig. 4.4. Generation of a misorientation band by splitting of an initial grain boundary disclination dipole

The total energy of the new split configuration (per unit length of disclinations) was written in [390] as the sum

$$W_2 = W_\alpha + W_\beta + W_{\alpha-\beta} + 2\gamma d - A \, , \tag{4.2}$$

where W_α and W_β are the elastic energies of the α- and β-dipoles (Fig. 4.4b), respectively, $W_{\alpha-\beta}$ is the energy of their interaction, γ the effective surface energy of the two new misorientation boundaries, and A the work by the external shear stress τ on the displacement d of the mobile β-dipole. The terms W_α and W_β are similar to that given by (4.1) with the replacement of ω by α and β, respectively. The energy of the dipole–dipole interaction $W_{\alpha-\beta}$ has been calculated as the work done to generate one dipole in the stress field of another dipole, thus resulting in [390]

$$W_{\alpha-\beta} = 2D\alpha\beta a^2 \left(\ln \frac{R^2}{4a^2 + d^2} - \frac{d^2}{4a^2} \ln \frac{4a^2 + d^2}{d^2} + 1 \right) . \tag{4.3}$$

The work A has been found in a similar way, giving $A = 2\tau\beta a d$.

To estimate γ, the authors [390] used the well-known approximation for the energy of a dislocation core [349] $W_c \approx Db^2/2$. For geometric reasons [349], the linear density ρ_g of 'geometrically necessary' dislocations [406] within a misorientation boundary is equal to β/b. However, one must also take into account the density of 'statistically stored' dislocations [406] which have different Burgers vector orientations and do not create any additional misorientation, but contribute to the effective surface energy of the boundary. Let the total dislocation density within the misorientation boundaries be $\rho_t = q\rho_g$, where $q \geq 1$ is a dimensionless parameter which accounts for the presence of statistically stored dislocations. Hence, the number of dislocation cores, N, within a misorientation boundary is estimated as $N = \rho_t d = q\beta d/b$, and the total core energy of a misorientation boundary is $NW_c \approx qD\beta bd/2$. This leads to the approximation $\gamma \approx NW_c/d \approx qD\beta b/2$ [390].

Let us now consider the energy difference [390]

$$\begin{aligned}\Delta W &= W_2 - W_1 \\ &= D\beta a^2 \left\{4\tilde{d}\left(q\tilde{b} - \frac{\tau}{D}\right) - 2\alpha\left[(1+\tilde{d}^2)\ln(1+\tilde{d}^2) - \tilde{d}^2\ln\tilde{d}^2\right]\right\} ,\end{aligned} \tag{4.4}$$

introducing the dimensionless quantities $\tilde{d} = d/2a$ and $\tilde{b} = b/2a$. The dependence of ΔW on the normalized displacement $\tilde{d}$ is given in Fig. 4.5 for $\tilde{b} = 10^{-3}$, $\alpha = \pi/200$, and different values of the external shear stress τ and parameter q [390]. Depending on τ and q, the character of the curves $\Delta W(\tilde{d})$ changes drastically from a monotonic decrease (Fig. 4.5a) to a non-monotonic decrease (Fig. 4.5b). In the first case, when $q = 1$ and τ and $\tilde{d}$ are small enough (here $0 \leq \tau \leq 0.007D$ and $\tilde{d} < 1$), $\Delta W < 0$ for $\tilde{d} > 0$. This means there is no energy barrier for generation of an MB. In the second case, when $q = 10$ with the same values of τ and $\tilde{d}$, $\Delta W > 0$ for $\tilde{d} < \tilde{d}_c$, where $\tilde{d}_c$ is determined by the equation $\Delta W(\tilde{d} = \tilde{d}_c) = 0$, and $\Delta W < 0$ for $\tilde{d} > \tilde{d}_c$. This means that there is an energy barrier for the formation of an MB under such a small τ. Following [390], let us introduce a characteristic critical value τ_g, which is determined by the equation $\tilde{d}_c = \tilde{l}$, where $\tilde{l} = l/2a$ is the normalized spacing between the geometrically necessary dislocations creating the tilt misorientation angle β. If $\tau < \tau_g$, the generation of an MB nucleus (an initial GB disclination dipole plus one dislocation dipole joined to this dipole and localized at a distance $\tilde{l}$ from it) is energetically unfavorable. If $\tau > \tau_g$, it is favorable. For the situation illustrated in Fig. 4.5b, $\tilde{l} \approx 0.1$ was assumed and hence $\tau_g \approx 0.005D$ for $q = 10$.

Using (4.4), one can estimate τ_g analytically for the case $\tilde{d} \ll 1$ as follows [390]:

$$\tau_g \approx D\tilde{b}\left[q - \frac{\tilde{\alpha}}{1-\tilde{\alpha}}\left(\frac{1}{2} - \ln\frac{\tilde{b}/\omega}{1-\tilde{\alpha}}\right)\right] , \tag{4.5}$$

where $\tilde{\alpha} = \alpha/\omega$. The plots $\tau_g(\tilde{\alpha})$ are shown in Fig. 4.5c for $q = 10$ and various values of the initial dipole strength ω [390]. One can conclude that τ_g decreases when both $\tilde{\alpha}$ and ω increase, in accordance with physical intuition. When $q = 10$, the numerical estimate for τ_g gives values of order $G/1000$ to $G/400$. The lower limit fits well with typical external stresses at the end of Stage II of deformation curves for BCC and FCC metals [264], while the upper limit corresponds to the level of deforming stress observed in NCMs (see Chap. 2).

4.1.3.2 Splitting of an Initial Quadrupole-Like Disclination Structure at a Grain Boundary Kink

Consider now a model which seems to be more realistic than the previous one. It is a grain boundary kink with a quadrupole-like disclination structure (Fig. 4.6a) [390]. This configuration is characterized by the disclination strengths $-\alpha$, $+\gamma$, $-\omega$ and $+\theta$ which satisfy the equation $\alpha + \omega = \gamma + \theta$. As

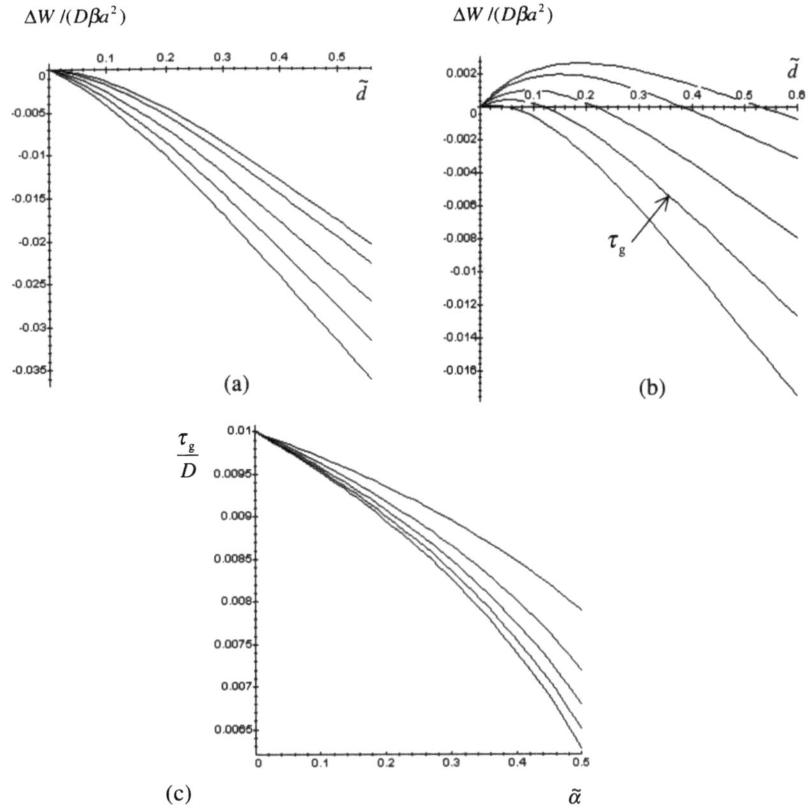

Fig. 4.5. Energy difference ΔW vs. the normalized misorientation band length $\tilde{d}$ (**a**, **b**) and critical shear stress τ_g/D vs. the normalized grain boundary disclination dipole strength $\tilde{\alpha}$ (**c**) for the model of a misorientation band being generated at a grain boundary disclination dipole. The plots $\Delta W(\tilde{d})$ are given for the external shear stress $\tau/D = 0$, 0.001, 0.003, 0.005, and 0.007 (*from top to bottom*) for the two different values of the parameter $q = 1$ (**a**) and 10 (**b**). The plots $\tau_g(\tilde{\alpha})/D$ are shown for the initial dipole strength $\omega = 0.01$, 0.02, 0.03, 0.04, and 0.05 (*from top to bottom*)

a result of the dislocation rearrangements which are suggested to be similar to those considered above, the quadrupole-like disclination structure issues a mobile two-axis dipole of partial wedge disclinations with strengths $\pm\beta$ and moving under the action of the grain boundary disclination stress field as well as an external shear stress τ (Fig. 4.6b) [390]. In a simple special case, the new split configuration has been suggested to consist of three two-axis disclination dipoles, one mobile (β-dipole) and two immobile (α- and θ-dipole). Once again, to find critical conditions for such a splitting transformation, the total energy W_1 of the initial quadrupole-like disclination structure and the energy W_2 of the new split configuration have been compared.

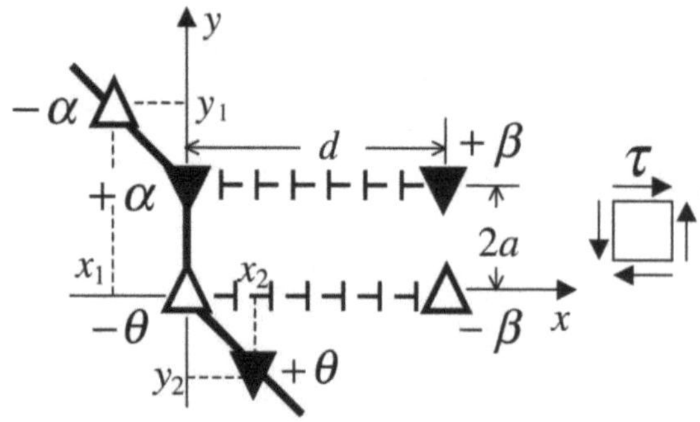

Fig. 4.6. Model of misorientation band generation by splitting of an initial quadrupole-like disclination configuration at a grain boundary kink

The energy difference $\Delta W = W_2 - W_1$ has been calculated as shown above with the result [390]

$$\Delta W = D\beta a^2 \Big\{ 4\tilde{d}\left(q\tilde{b} - \frac{\tau}{D}\right) - (\alpha + \theta)\left[(1+\tilde{d}^2)\ln(1+\tilde{d}^2) - \tilde{d}^2 \ln \tilde{d}^2\right] + \alpha\Psi(\tilde{x}_1, \tilde{y}_1, \tilde{d}) - \theta\,\Psi(\tilde{x}_2, \tilde{y}_2, \tilde{d}) \Big\}, \quad (4.6)$$

where

$$\begin{aligned}\Psi(x,y,z) = &-(x^2+y^2)\ln(x^2+y^2) + \left[x^2+(y-1)^2\right]\ln\left[x^2+(y-1)^2\right] \\ &+\left[(x-z)^2+y^2\right]\ln\left[(x-z)^2+y^2\right] \\ &-\left[(x-z)^2+(y-1)^2\right]\ln\left[(x-z)^2+(y-1)^2\right], \end{aligned} \quad (4.7)$$

$\tilde{x}_i = x_i/2a$ and $\tilde{y}_i = y_i/2a$, $i = 1, 2$.

The curves $\Delta W(\tilde{d})$ are given in Fig. 4.7 for $\tilde{b} = 10^{-3}$, $\alpha = \pi/200$, $\theta = \pi/300$, $\tilde{x}_1 = -\tilde{x}_2 = \tilde{y}_2 = -1$, $\tilde{y}_1 = 2$ and different values of τ and q [390]. Depending on τ and q, the curves $\Delta W(\tilde{d})$ behave in quite different ways. When $q \leq 5$ (e.g., see Fig. 4.7a for $q = 1$), they are similar to those considered in the previous section. When $q \geq 7$ (e.g., see Fig. 4.7b for $q = 10$), some of them increase monotonically. One can conclude that, for the given values of the parameters, the critical external shear stress τ_g is higher than for the case of grain boundary disclination dipole splitting (Sect. 4.1.3.1).

Introducing (4.6) with (4.7) into the equation $\Delta W(\tilde{d} = \tilde{l}) = 0$ for $\tilde{d} \ll 1$, the authors obtained [390]

$$\frac{\tau_g}{D} \approx \tilde{b}q - \frac{\tilde{b}(\tilde{\alpha}+\tilde{\theta})}{4(1-\tilde{\alpha}-\tilde{\theta})}\left(1 - 2\ln\frac{\tilde{b}/\omega}{1-\tilde{\alpha}-\tilde{\theta}}\right) \quad (4.8)$$

$$+(1-\tilde{\alpha}-\tilde{\theta})\frac{\omega^2}{4\tilde{b}}\left[\tilde{\alpha}\Psi\left(\tilde{x}_1, \tilde{y}_1, \tilde{d} = \frac{\tilde{b}/\omega}{1-\tilde{\alpha}-\tilde{\theta}}\right) - \tilde{\theta}\Psi\left(\tilde{x}_2, \tilde{y}_2, \tilde{d} = \frac{\tilde{b}/\omega}{1-\tilde{\alpha}-\tilde{\theta}}\right)\right],$$

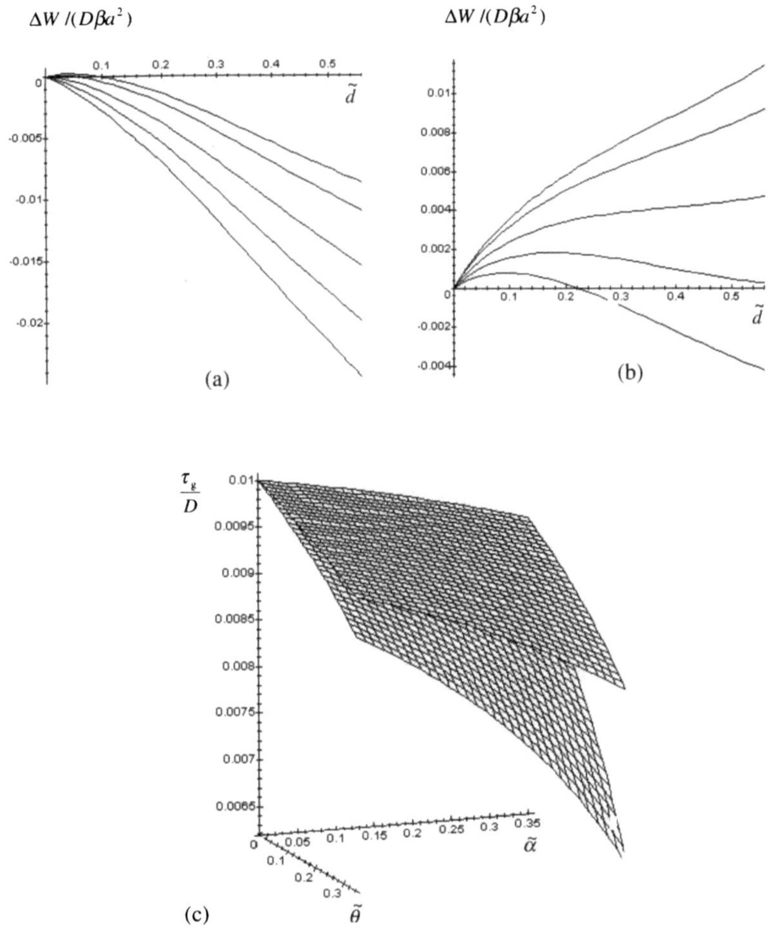

Fig. 4.7. Energy difference ΔW vs. the normalized misorientation band length $\tilde{d}$ (**a, b**) and critical shear stress τ_g/D vs. the normalized grain boundary disclination dipole strengths $\tilde{\alpha}$ and $\tilde{\theta}$ (**c**) for the model of a misorientation band being issued by a quadrupole-like disclination configuration localized at a grain boundary kink. The plots $\Delta W(\tilde{d})$ are given for the external shear stress $\tau/D = 0$, 0.001, 0.003, 0.005, and 0.007 (*from top to bottom*) for the two different values of the parameter $q = 1$ (**a**) and 10 (**b**). The plots $\tau_g(\tilde{\alpha}, \tilde{\theta})/D$ are shown for the initial disclination strength $\omega = 0.01$ (*upper surface*) and 0.05 (*lower surface*)

where $\tilde{\alpha} = \alpha/\omega$ and $\tilde{\theta} = \theta/\omega$. Three-dimensional plots $\tau_g(\tilde{\alpha}, \tilde{\theta})$ are shown in Fig. 4.7c for $q = 10$, $\tilde{x}_1 = -\tilde{x}_2 = \tilde{y}_2 = -1$, $\tilde{y}_1 = 2$ and two values of the initial disclination strength $\omega = \alpha + \beta + \theta$, $\omega = 0.01$ and 0.05 [390]. It is seen that τ_g decreases when $\tilde{\alpha}$, $\tilde{\theta}$ and ω increase. Once again, this is in accordance with intuition. When $q = 10$, the numerical estimate for τ_g once again gives values of order $G/1000$ to $G/400$.

4.1.3.3 Splitting of an Initial Quadrupole-Like Disclination Structure at a Grain Boundary Triple Junction

Another more realistic model is a grain boundary triple junction having a quadrupole-like disclination structure with the geometry shown in Fig. 4.8 with $d = 0$ [390]. Again this disclination configuration issues a mobile two-axis dipole of partial wedge disclinations with strengths $\pm\beta$ and moving under the action of the grain boundary disclination stress field and external shear stress τ. It has been suggested that the new split configuration consists of four two-axis disclination dipoles, one mobile (β-dipole) and three immobile (α-, θ- and γ-dipole). These three immobile dipoles are formed by the 'central' negative triple-junction disclination with strength $-\omega$ and three positive disclinations with strengths $+\alpha$, $+\theta$ and $+\gamma$, localized at the joined grain boundaries. The relation $\omega = \alpha + \theta + \gamma$ must be valid.

The calculation procedure for the difference in the total energies of the initial quadrupole-like disclination structure and new split configuration is absolutely parallel to that described in Sect. 4.1.3.1. The final result is [390]

$$\Delta W = D\beta a^2 \Bigg\{ 4\tilde{d}\left(q\tilde{b} - \frac{\tau}{D}\right) - (2\alpha + \theta + \gamma)\left[(1+\tilde{d}^2)\ln(1+\tilde{d}^2) - \tilde{d}^2 \ln \tilde{d}^2\right] \\ -\theta\Psi(\tilde{x}_1, \tilde{y}_1, \tilde{d}) - \gamma\Psi(\tilde{x}_2, \tilde{y}_2, \tilde{d}) \Bigg\}, \tag{4.9}$$

where the function Ψ is given by (4.7). The dependences $\Delta W(\tilde{d})$ are plotted in Fig. 4.9 for $\tilde{b} = 10^{-3}$, $\alpha = \pi/300$, $\theta = \pi/100$, $\gamma = \pi/200$, $\tilde{x}_1 = -\tilde{x}_2 = \tilde{y}_1 = \tilde{y}_2 = -1$, and different values of τ and q [390]. This set of parameters gives curves very similar to those considered in Sect. 4.1.3.1 (Fig. 4.5) with close values for the critical external stress τ_g when $q \geq 5$. Comparing these plots with those in Fig. 4.7, one can conclude that both $\Delta W(\tilde{d})$ and τ_g depend on

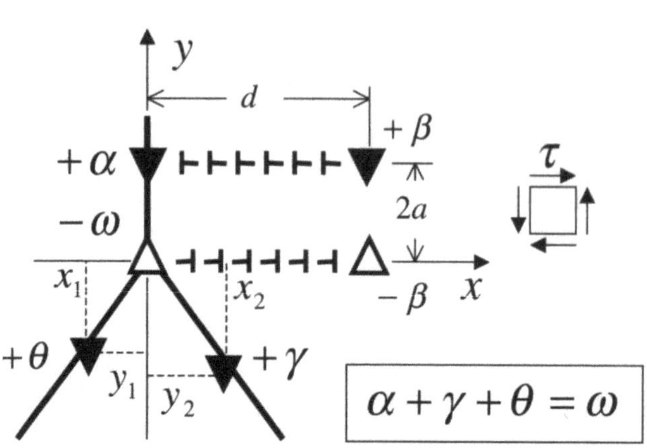

Fig. 4.8. Model of misorientation band generation by splitting of an initial quadrupole-like disclination configuration at a triple junction of grain boundaries

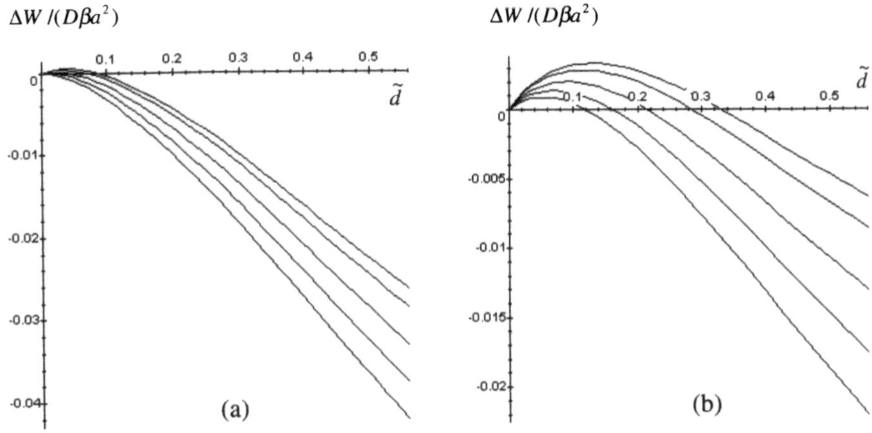

Fig. 4.9. Energy difference ΔW vs. the normalized length d of a misorientation band being issued by a quadrupole-like disclination configuration localized at a triple junction of grain boundaries under the external shear stress $\tau/D = 0$, 0.001, 0.003, 0.005, and 0.007 (*from top to bottom*) for the two different values of the parameter $q = 1$ (**a**) and 10 (**b**)

the specific arrangement of the disclinations within the initial quadrupole-like disclination structure.

Substituting (4.9) with (4.7) into the equation $\Delta W(\tilde{d} = \tilde{l}) = 0$ gives the following result for $\tilde{d} \ll 1$ [390]:

$$\begin{aligned}\frac{\tau_g}{D} \approx \tilde{b}q - \frac{\tilde{b}(2\tilde{\alpha} + \tilde{\theta} + \tilde{\gamma})}{4(1 - \tilde{\alpha} - \tilde{\theta} - \tilde{\gamma})}\left(1 - 2\ln\frac{\tilde{b}/\Omega}{1 - \tilde{\alpha} - \tilde{\theta} - \tilde{\gamma}}\right) \\ -(1 - \tilde{\alpha} - \tilde{\theta} - \tilde{\gamma})\frac{\Omega^2}{4\tilde{b}}\left[\tilde{\theta}\Psi\left(\tilde{x}_1, \tilde{y}_1, \tilde{d} = \frac{\tilde{b}/\Omega}{1 - \tilde{\alpha} - \tilde{\theta} - \tilde{\gamma}}\right)\right. \\ \left. + \tilde{\gamma}\Psi\left(\tilde{x}_2, \tilde{y}_2, \tilde{d} = \frac{\tilde{b}/\Omega}{1 - \tilde{\alpha} - \tilde{\theta} - \tilde{\gamma}}\right)\right],\end{aligned} \quad (4.10)$$

where $\tilde{\alpha} = \alpha/\Omega$, $\tilde{\theta} = \theta/\Omega$, $\tilde{\gamma} = \gamma/\Omega$, and $\Omega = \alpha + \beta + \theta + \gamma$ is the strength of the initial quadrupole-like disclination structure.

Using (4.10), the authors [390] have investigated three-dimensional plots $\tau_g(\tilde{\alpha}, \tilde{\theta})$ for different constant values of $\tilde{\gamma}$, $\tau_g(\tilde{\alpha}, \tilde{\gamma})$ for different constant values of $\tilde{\theta}$, and $\tau_g(\tilde{\theta}, \tilde{\gamma})$ for different constant values of $\tilde{\alpha}$ at the following parameter values: $q = 5$, $\tilde{x}_1 = -\tilde{x}_2 = \tilde{y}_1 = \tilde{y}_2 = -1$, and two values of the initial disclination strength Ω. All the plots turn out to be very similar to those shown in Fig. 4.7c and hence we do not represent them here. They demonstrate that τ_g decreases when $\tilde{\alpha}$, $\tilde{\theta}$, $\tilde{\gamma}$ and Ω increase [390]. When $q = 5$–10, the numerical estimate for τ_g once again gives values of order $G/1000$ to $G/400$.

4.1.4 Regimes of Misorientation Band Propagation

In the previous sections, we have considered three different models of MB generation by analysing the energy expressions (4.4), (4.6) and (4.9) for the limit $\tilde{d} \ll 1$. In order to study the further propagation of the MB, the authors [390] used the same expressions but for $\tilde{d} \geq 1$. Consider again the simplest model, i.e., the splitting of an initial grain boundary disclination dipole (see Sect. 4.1.3.1, Fig. 4.4).

A characteristic example of the graphical representation of (4.4) for the case $\tilde{d} \geq 1$ is given in Fig. 4.10 for $\alpha = \pi/200$, $\tilde{b} = 10^{-3}$, $q = 3$, and different values of the external shear stress τ [390]. Depending on the value of τ, the curves $\Delta W(\tilde{d})$ behave in different ways. When τ is smaller than some limiting quantity τ_{p} (e.g., $\tau_{\mathrm{p}} \approx 0.003D$ for $q = 3$), the curves $\Delta W(\tilde{d})$ are non-monotonic and achieve their minima, which determine equilibrium values of the MB length $\tilde{d}_{\mathrm{eq}}$. When $\tau > \tau_{\mathrm{p}}$, the curves $\Delta W(\tilde{d})$ decrease monotonically and $\tilde{d}_{\mathrm{eq}}$ is absent.

To find an analytical estimate for τ_{p}, one can solve the equation $\Delta W(\tau) = 0$ for the limiting case $\tilde{d} \to \infty$. This gives [390]

$$\tau_{\mathrm{p}} = Dq\tilde{b} \,. \tag{4.11}$$

Obviously, this provides the linear relation between τ_{p} and q, where q characterizes the effective surface energy of the misorientation boundaries.

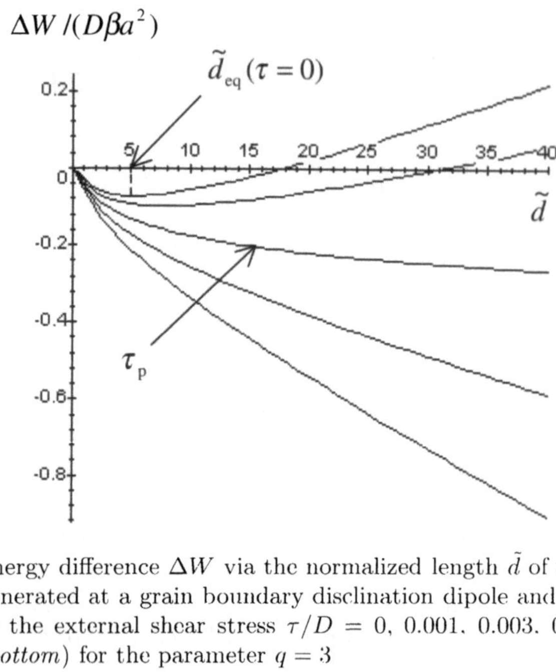

Fig. 4.10. Energy difference ΔW via the normalized length $\tilde{d}$ of a misorientation band being generated at a grain boundary disclination dipole and propagating far from it under the external shear stress $\tau/D = 0$, 0.001, 0.003, 0.005, and 0.007 (*from top to bottom*) for the parameter $q = 3$

The equilibrium length $\tilde{d}_{\rm eq}$ is found from the standard equation $\partial\Delta W/\partial\tilde{d} = 0$ with $\partial^2\Delta W/\partial\tilde{d}^2 > 0$ and $\tilde{d} > 1$. The final result reads [390]

$$\tilde{d}_{\rm eq} \approx \frac{\alpha}{q\tilde{b} - \tau/D}\,, \tag{4.12}$$

where $\tau < \tau_{\rm p}$. The dependence of $\tilde{d}_{\rm eq}$ on τ is illustrated in Fig. 4.11 for $\alpha = \pi/200$, $\tilde{b} = 10^{-3}$, and different values of q. One can see that $\tilde{d}_{\rm eq}$ increases with increasing τ and decreasing q.

It has thus been shown in [390] that, depending on the external shear stress τ, two main regimes of MB propagation are possible: stable and unstable propagation. When $\tau < \tau_{\rm p}$, the MB propagation is stable and may be characterized by the equilibrium length $\tilde{d}_{\rm eq}$. When $\tau > \tau_{\rm p}$, the MB propagation is unstable and there is no equilibrium length. In reality, this means that the MB will propagate until it meets an obstacle like a grain boundary or another MB (or some other defect configurations playing the role of obstacles) in which case the question of its further propagation must be considered again.

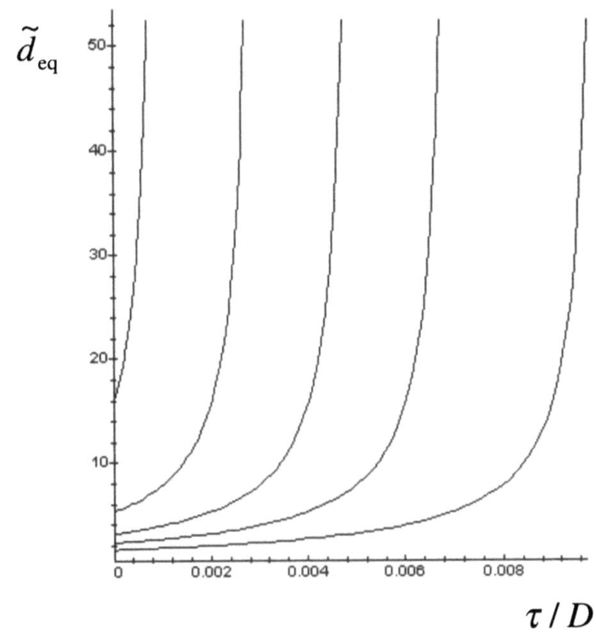

Fig. 4.11. Normalized equilibrium length $\tilde{d}_{\rm eq}$ of a misorientation band via the external shear stress τ/D for different values of the parameter q = 1, 3, 5, 7, and 10 (*from top to bottom*)

4.1.5 Computer Simulation of Dislocation–Disclination Interactions

To check and refine the models of MB propagation through an ensemble of edge dislocations and calculate some important parameters of dislocation–disclination interactions such as the effective length of dislocation capturing by a disclination dipole (this length was treated as the distance from a partial disclination line at which the corresponding edge dislocation must be stopped to provide the conservative motion 'ahead' of the partial disclination), the method of 2D dislocation–disclination dynamics has been used in [396]. This approach and some results have also been quoted in [390].

The computer code objects are straight edge dislocations and straight wedge disclinations which can move within a two-dimensional rectangular box of an infinite elastically isotropic medium (Fig. 4.12). Periodic boundary conditions are realized. The box sizes are chosen as 1×1 mm^2. Defect lines are normal to the box plane. Dislocations are characterized by their Burgers vectors b_x or b_y, coordinates $(x^{(i)}, y^{(i)})$ and velocities $(\dot{x}^{(i)}, \dot{y}^{(i)})$, where $i = 1, \ldots, n$ and n is the number of dislocations. All disclinations are arranged in dipole configurations which are assumed to be immobile and considered as sources of elastic fields. The dipoles are characterized by their strengths $\omega^{(j)}$, by the size and orientation of their arms, and by the coordinates of the arm central points $(X^{(j)}, Y^{(j)})$, where $j = 1, \ldots, N$ and N is the number of disclination dipoles.

In such a computer model [396], dislocations can move by gliding or climbing under the action of the total force due to external loading, elastic fields

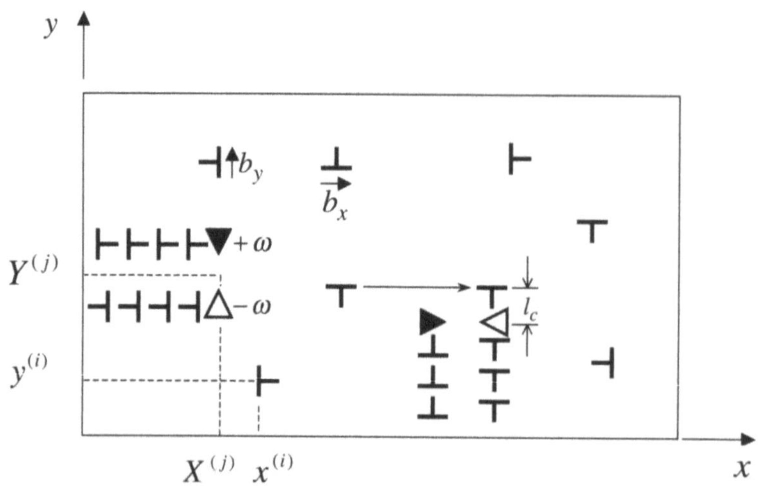

Fig. 4.12. 2D box for computer simulation of a dislocation–disclination ensemble. The parameter l_c denotes the length of dislocation capturing by the disclination dipole

of other defects and dynamic friction. The dislocation dynamics is than governed by Newton's law

$$m^{(i)}\ddot{x}^{(i)} = F_x^{(i)} , \tag{4.13}$$
$$m^{(i)}\ddot{y}^{(i)} = F_y^{(i)} , \tag{4.14}$$

where $m^{(i)}$ is the effective mass of the ith dislocation, and $\ddot{x}^{(i)}$ and $\ddot{y}^{(i)}$ are the x- and y-components of its acceleration, respectively. The p-component ($p = x, y$) of the resulting force on the ith dislocation, denoted $F_p^{(i)}$, is assumed to be a superposition

$$F_p^{(i)} = F_p^{\mathrm{def}(i)} + F_p^{\mathrm{fr}(i)} + F_p^{\mathrm{ext}(i)} , \tag{4.15}$$

where $F_p^{\mathrm{def}(i)} = e_{mpl}\sigma_{lk}^{(i)} b_k^{(i)} s_m^{(i)}$ is the elastic force due to all other defects, $F_p^{\mathrm{fr}(i)} = -\tau(v)b_p^{(i)}$ the dynamic friction force, and $F_p^{\mathrm{ext}(i)} = \tau^{\mathrm{ext}} b_p^{(i)}$ the external driving force. Here e_{mpl} denotes the permutation symbol, $\sigma_{lk}^{(i)}$ is the resulting elastic stress due to all other defects as measured at the point where the ith dislocation is located, $b_k^{(i)}$ is the k-component of its Burgers vector, and $s_m^{(i)}$ is the m-component of the unit vector tangent to the dislocation line. All indexes p, m, l and k may denote x- or y-components. The shear stress $\tau(v)$ characterizes the crystalline lattice friction and depends on the dislocation velocity v. The external stress τ^{ext} is created by an external load applied to the solid.

Given the initial coordinates and velocities of every defect, the system of equations of motion (4.13)–(4.14) with (4.15) has been solved numerically to find the dependences of the coordinates x, y and the velocity components v_x and v_y on the time t.

The above approach has been used to consider the elastic interaction of a gliding edge dislocation with a two-axis wedge disclination dipole in pure copper [396]. The magnitude of the Burgers vector has been taken as $b_x = 0.256$ nm, and hence the corresponding dislocation mass (per unit length of the dislocation) follows as $m = \rho b_x^2/2 \approx 2.9 \times 10^{-16}$ kg m^{-1} (see [407], p. 73). The position of the disclination dipole has been fixed at the central point ($X = 500$ μm, $Y = 500$ μm) of the simulation box. In obtaining the following results, the force F_p^{ex} has been neglected to catch the main features of elastic dislocation–disclination interactions as they are. The elastic force $F_x^{\mathrm{def}} = \sigma_{xy} b_x$ has been calculated with the disclination elastic stress field σ_{xy} taken from [56]. The dynamic friction force F_p^{fr} has been taken as $F_x^{\mathrm{fr}}(t) = -Bv(t)$, where $B = 1.7 \times 10^{-5}$ Pa s is characteristic for pure copper [408] (see also [407], p. 76).

Some typical situations have been studied for different orientations of the dipole arm, initial positions and velocities of the dislocation [390, 396]. It has been shown that the dislocation behavior may vary strongly depending on the problem parameters. However, it is governed mainly by the elastic stress field of the disclination dipole.

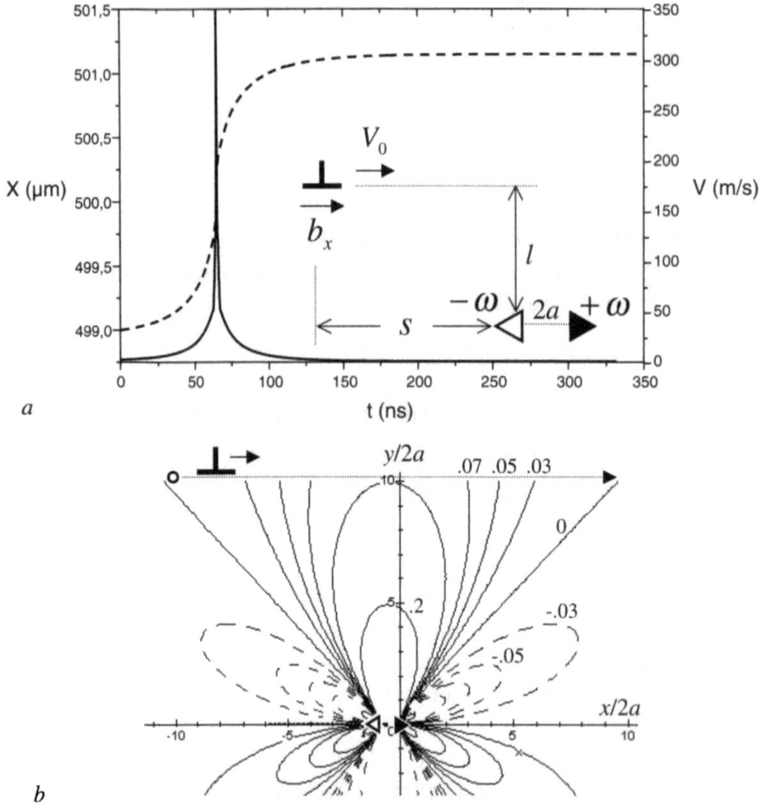

Fig. 4.13. Accelerated glide of an edge dislocation along the arm of a disclination dipole. The *dashed* and *continuous curves* in (**a**) represent the dislocation position $x(t)$ and velocity $v(t)$, respectively, when the dislocation glides in the field of the dipole positive long-range shear stress σ_{xy} (**b**). The calculations have been carried out for the following parameter values: $\omega = 0.01$, $b_x = 0.256$ nm, $2a = 100$ nm, $s = 1000$ nm, $l = 1100$ nm, $x_0 = 499$ μm, and $v_0 = 0.01$ m s^{-1}

For example, consider a disclination dipole with strength $\omega = 0.01$ and arm $2a = 100$ nm. Let the dislocation move with initial velocity $v_0 = 0.01$ m s^{-1} quite far from the disclination dipole, in the manner shown in Fig. 4.13 [390,396]. The corresponding plots for the dislocation coordinate $x(t)$ (dashed curve) and velocity $v(t) = \dot{x}(t)$ (continuous curve) are given in Fig. 4.13a. Figure 4.13b shows the distribution of the dipole shear stress σ_{xy} in units of $D\omega$. The empty and black circles in Fig. 4.13b schematically denote the initial and final dislocation positions, respectively. In fact, the dislocation starts to glide when under the action of the positive shear stress σ_{xy} of the disclination dipole. (The initial dislocation position is at the upper left corner in Fig. 4.13b.) As a result, the dislocation glides with an acceleration and its velocity becomes very high when it passes over the disclination dipole (Fig. 4.13a), in the field of the strongest disclination stresses (the upper central region in Fig. 4.13b). At the same time, the friction force also achieves it maximum value. Therefore, when the dislocation has passed the region of maximum stress values over the dipole, its velocity starts to fall quickly. As a result, the dislocation moves away from the dipole with a negative acceleration until the point where the dipole stress goes to zero (the upper right region in Fig. 4.13b). One can see that there is no effect of dislocation capturing by the disclination dipole in this case.

The question arises as regards the problem conditions for which dislocation capturing by the disclination dipole would be possible for a given defect configuration. Calculations [390,396] have demonstrated that dislocation capturing is only possible when its initial position is just over the center of the disclination dipole, at a very small distance from it, and the disclination strength is very small. Hence, let the dislocation begin its motion just near a disclination dipole with ten times lower strength $\omega = 0.001$ and the same arm (Fig. 4.14). The dislocation is accelerated from $v_0 = 0$ within the region of the relatively higher positive shear stresses of the dipole. (The initial dislocation position is at the central part of Fig. 4.14b.) It is then stopped at the zero stress contour above the dipole, just near the positive disclination (see Fig. 4.14b). This means that the disclination dipole has captured the dislocation. However, capturing occurs only at very small distances from the dipole. This distance (here $l = 1$ nm) is much smaller than the spacing $l_c \approx b_x/\omega = 256$ nm between dislocations in the low-angle tilt walls whose edges are described by the disclination dipole (see Fig. 4.12). Therefore, such a small capturing length cannot provide the mechanism of conservative motion of a disclination dipole in a direction normal to its arm, by capturing or issuing edge dislocations.

Computer simulations [390, 396] show that the dynamics of the edge dislocation are totally governed by the elastic field of the disclination dipole. The dislocation is accelerated when it appears in the region of increasing disclination stress, while further away, when this field decreases, the dislocation is hampered by the force of dynamic friction and always stopped at the line of zero-level disclination shear stress. Hence, the dislocation behavior is determined by its initial position with respect to the disclination dipole and does not in fact depend on its initial velocity (at least for those velocity values which have been used in simulations). The computer model [390, 396] has confirmed that the two-axis dipole of wedge disclinations can move conservatively along the direction parallel to the dipole arm by capturing edge dislocations. However, the dipole motion along the normal to its arm cannot be explained correctly within the existing theoretical models and needs further investigation.

On the basis of the results of the theoretical models for MB generation and propagation considered in Sect. 4.1, one can draw the following conclusions:

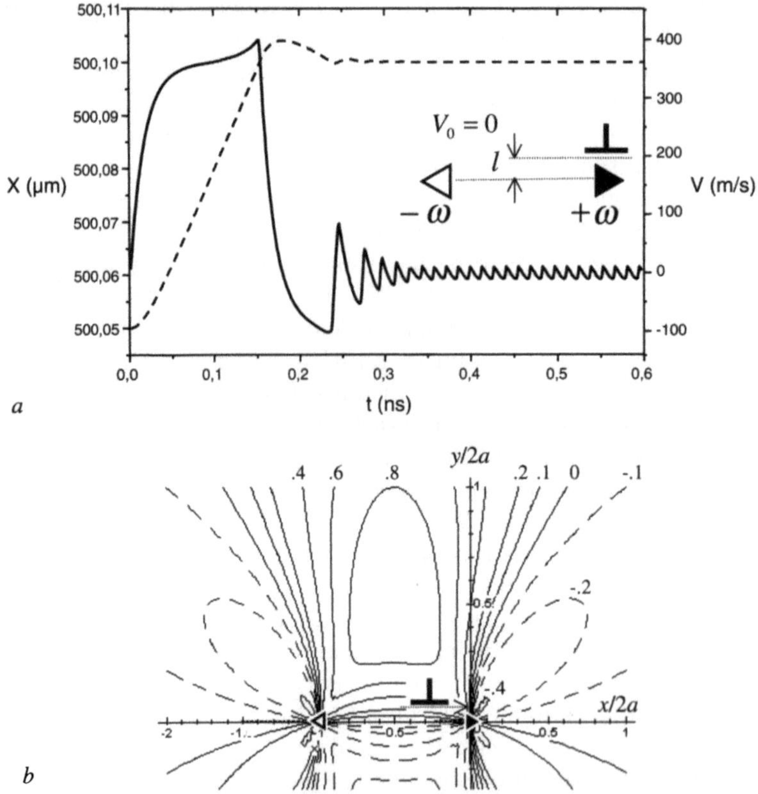

Fig. 4.14. Accelerated glide of an edge dislocation along the arm of a disclination dipole and capture of the dislocation by the dipole. The *dashed* and *continuous curves* in plot (**a**) represent the dislocation position $x(t)$ and velocity $v(t)$, respectively, when the dislocation glides in the field of the dipole positive short-range shear stress σ_{xy} (**b**). The calculations have been carried out for the following parameter values: $\omega = 0.001$, $b_x = 0.256$ nm, $2a = 100$ nm, $s = 50$ nm, $l = 1$ nm, $x_0 = 500.05$ μm, and $v_0 = 0$

- Disclination models of MB generation at grain boundary faults like kinks and grain boundary junctions predict the existence of a critical external shear stress τ_g which is necessary for MB generation events to take place. The numerical estimate for the critical stress gives values of order $G/1000$ to $G/400$. The lower limit is in a good agreement with typical external stresses at the end of Stage II of deformation curves for conventional polycrystalline BCC and FCC metals, while the upper limit corresponds to the level of deforming stress observed in NCMs.
- The critical shear stress τ_g depends strongly on the geometry and strengths of initial grain boundary disclination configurations, the misorientation angle, and the effective surface energy of arising misorientation bound-

aries. It increases when the initial disclination strength decreases and the misorientation angle increases. The critical stress varies in direct proportion to the effective surface energy of the misorientation boundaries.

- Disclination models of MB propagation predict the existence of a limiting external shear stress $\tau_{\rm p}$, which separates two main regimes of MB propagation: stable and unstable propagation. When the external stress is lower than the limiting stress, the MB propagation is stable and may be characterized by the equilibrium MB length which increases when the external stress increases and the effective surface energy of misorientation boundaries decreases. If the external stress is higher than the limiting stress, the MB propagation is unstable. The limiting external stress varies in direct proportion to the effective surface energy of misorientation boundaries.
- Computer simulations by means of a 2D dislocation–disclination dynamics code have shown that the existing models of disclination dipole motion must be reconsidered, taking into account the conclusion that the disclination dipole cannot move conservatively along the normal to its arm by capturing edge dislocations.

4.2 Motion of Grain Boundary Disclinations

Transformations of grain boundaries often strongly influence both the structure and properties of polycrystalline and nanocrystalline materials [56, 210, 227, 326–330, 339, 409–416]. In particular, changes in the misorientation parameters of grain boundaries capable of resulting in grain rotations have been detected experimentally in polycrystalline and nanocrystalline materials under (super)plastic deformation [326–330, 409]. Ke et al. [189, 196] observed in situ that plastic deformation of nanocrystalline gold films with grain size $d < 25$ nm occurred through grain boundary sliding and grain rotation near the tips of opening cracks. Noskova and Volkova [413–415] have also reported in situ TEM observation of grain boundary sliding and grain rotation in nanocrystalline pure metals (Cu, Ni and Ti with grain size $d =$ 20–40 nm) and $Fe_{73.5}Cu_1Nb_3Si_{13.5}B_9$ alloy (with grain size $d \approx 10$ nm) under active unidirectional tension with strain rate 10^{-5} s^{-1}. It is important that rotation of grains does not always require application of an external mechanical load. It is sometimes enough to carry out a special thermal treatment. For example, grain rotations have been observed experimentally in thin films of gold under thermal treatment [416].

According to contemporary theoretical representations of grain boundaries, changes in their misorientation parameters occur via motion of grain boundary disclinations [56, 263–266, 417]. Such disclinations are intensively generated in polycrystals and NCMs under highly nonequilibrium fabrication conditions [15]. These conditions are typical for the technologies of severe plastic deformation and ball milling that give submicro- and nanostructured materials with nonequilibrium grain boundaries containing grain boundary and triple junction disclinations [28,197,217,218,405] (see also Chap. 2). It is quite natural that in such NCMs, as in conventional polycrystals under large deformations, the processes of rotational deformation are assumed to be realized by means of conservative motion of disclination dipoles [390,418–420].

Motion of grain boundary disclinations in plastically deformed materials is commonly treated as being associated with absorption of lattice dislocations (generated and moving in grains under the action of a mechanical load) by grain boundaries [56,263,409]. According to [409], this micromechanism is responsible for experimentally observed grain rotations in fine-grained materials during (super)plastic deformation. However, the subsequent motion of a grain boundary disclination along a grain boundary requires processes of dislocation absorption to be well ordered in space and time. In particular, lattice dislocations with certain Burgers vectors have to reach the grain boundary only in the vicinity of the disclination that moves along the boundary due to acts of absorption of these dislocations. This is in evident contradiction with the fact that sources of lattice dislocations in plastically deformed materials are commonly distributed in a rather irregular way within a grain and are not therefore capable of providing the regular flow of dislocations to the disclination moving along a grain boundary. Moreover, in the situation with grain rotations in thin films of gold under thermal treatment [416], absorption of lattice dislocations by grain boundaries hardly plays an important role in changes in grain boundary misorientation parameters, because the dislocation density in grain interiors is too low to cause grain rotations.

On the basis of these conclusions, alternative theoretical models [419–422] have been proposed. These models have used the idea that grain boundary disclinations can move along grain boundaries due to emission (in contrast to absorption [409]) of lattice dislocations from grain boundaries into adjacent grain interiors. It has also been assumed that the grain boundary disclinations are partial, i.e., they may be represented as ragged walls of edge dislocations [56,263,397,398] (this assumption is always valid in practice), and accordingly of wedge-type. In this case, a grain boundary disclination can emit a pair of edge lattice dislocations with Burgers vectors corresponding to the glide systems in the adjacent grains. The sum of these lattice Burgers vectors is supposed to be equal to the Burgers vector of a grain boundary dislocation from the ragged wall. (Strictly speaking, this condition is not necessary and has been used for the sake of simplicity. One could also take into consideration a difference grain boundary dislocation whose Burgers vector would compensate a possible disparity.) The emission of dislocation pairs allows the grain boundary disclination to move conservatively along the grain boundary, and the events of dislocation emission are determined by the conditions at the grain boundary disclination. We consider below the main results of the models [419–422] which describe the rotational mode of plastic deformation in fine-grained materials as being realized through conservative motion of dipoles of grain boundary disclinations emitting pairs of lattice dislocations.

4.2.1 Changes in Grain Boundary Misorientation

Following the theory of grain boundaries, two fragments of a grain boundary that are characterized by different values of the tilt misorientation parameter are divided by the line of a grain boundary disclination [56, 263, 265, 417]. More precisely, the line that separates the two boundary fragments with tilt misorientation parameters θ_1 and θ_2, respectively, is described as the line of a grain boundary wedge disclination with strength $\omega = \theta_1 - \theta_2$ (Fig. 4.15). In the framework of the representations discussed, the evolution (in time) of the tilt misorientation along grain boundaries is treated as being related to the motion of disclinations.

In the framework of the model [421, 422], an elementary act of transfer (through a distance l) of a grain boundary disclination with strength ω is accompanied by emission of two lattice dislocations with Burgers vectors $\boldsymbol{b}_1$ and $\boldsymbol{b}_2$ from the grain boundary into the adjacent grains I and II, respectively (Fig. 4.16). The disclination with strength ω can be treated as terminating a ragged wall of periodically spaced grain boundary dislocations with identical Burgers vectors $\boldsymbol{b}$ and spacing (period) l. This dislocation representation is relevant to both small- and large-angle grain boundaries with grain boundary dislocations having a 'large' crystal lattice Burgers vector in the case of small-angle grain boundaries and a 'small' displacement-shift-complete(DSC)-lattice Burgers vector in the case of large-angle grain boundaries [423, 424]. Assuming that the spacing l between such dislocations is the distance of an elementary transfer of the disclination, the Burgers vectors of the grain boundary dislocations and strength ω of the GB disclination obey the equations $\boldsymbol{b}_1 + \boldsymbol{b}_2 = \boldsymbol{b}$ and $|\boldsymbol{b}| = b \approx l\omega$. From the former equation, one finds that the magnitude b of the Burgers vector bears the following relationship with the magnitudes b_1 and b_2 of the Burgers vectors of the emitted dislocations:

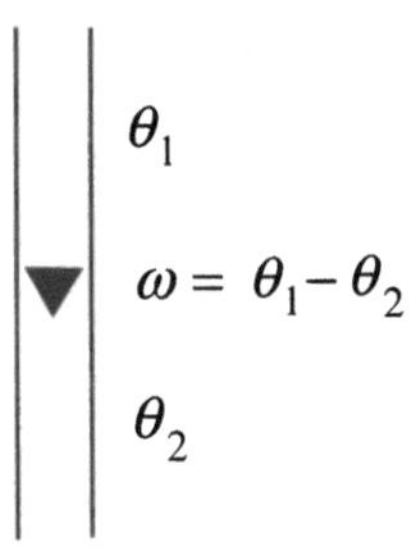

Fig. 4.15. A grain boundary disclination (*black triangle*) separates boundary fragments characterized by different values θ_1 and θ_2 of the tilt misorientation

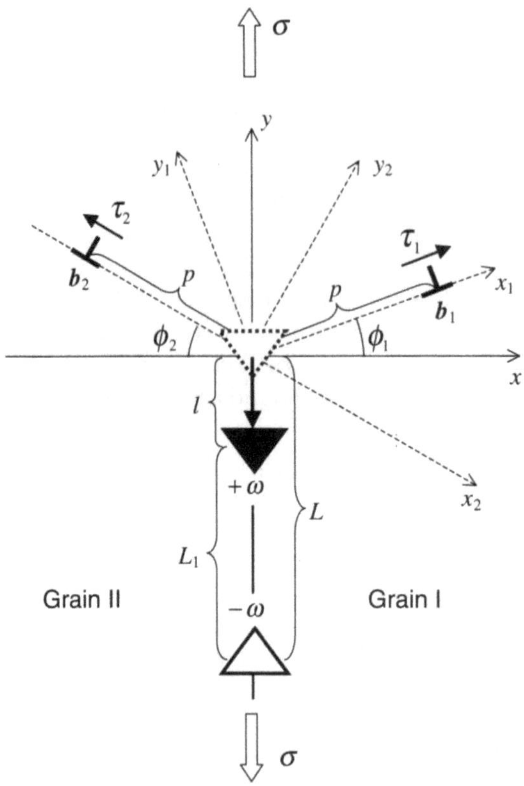

Fig. 4.16. Displacement of the wedge disclination (*black triangle*) with strength ω from its initial position (*dashed triangle*) by the distance l is accompanied by the emission of two lattice dislocations with Burgers vectors $\boldsymbol{b}_1$ and $\boldsymbol{b}_2$. The ω-disclination moves along the grain boundary plane (which is perpendicular to the figure plane and intersects it along the y-axis) towards another disclination (*white triangle*) with strength $-\omega$. The x-axis is normal to the grain boundary plane. The x_1y_1 and x_2y_2 coordinate systems are associated with the gliding planes of emitted dislocations. ϕ_1 and ϕ_2 are the angles between the normal to the grain boundary plane and the gliding planes of the first and the second dislocations, respectively. τ_1 and τ_2 are the shear stresses acting along the gliding planes of the first and second dislocations, respectively. L and L_1 are the distances between the disclinations before and after displacement of the ω-disclination

$$b = b_1\left(\cos\phi_1 + \frac{b_2}{b_1}\cos\phi_2\right) = b_1\frac{\sin(\phi_1+\phi_2)}{\sin\phi_2}\,, \tag{4.16}$$

where ϕ_1 and ϕ_2 are the angles between the normal to the grain boundary plane and the gliding planes of the first and second emitted lattice dislocations, respectively (Fig. 4.16). Notice that $b < b_1, b_2$ in the case of large-angle grain boundaries (containing dislocations with DSC-lattice Burgers vectors).

The consequent emission of lattice dislocation pairs (Fig. 4.16) causes a change in the grain boundary misorientation along large fragments of the grain boundary plane. This process is capable of giving rise to the rotation of a grain as a whole.

The above model is approximate. First, the authors [421, 422] restricted their consideration to the z-independent situation with a disclination dipole at a tilt boundary characterized by two macroscopic geometric parameters, the angles ϕ_1 and ϕ_2 (Fig. 4.16). However, tilt boundaries which are effectively described in terms of disclinations [56, 263] represent the most widespread type of grain boundary in real materials, so that this model covers most real grain boundaries.

Second, the choice of the disclination dipole as a subject for theoretical analysis (addressing the grain boundary disclination motion) has been related in this model to the two following features:

- the long-range stress field of the moving disclination should be screened,
- there is experimental evidence (see reviews in [56,263–266] and references therein) that disclinations form dipole, quadrupole and multipole configurations in real materials under large deformations.

In principle, one could consider an individual grain boundary disclination whose elastic fields would be screened by outer boundaries of the solid, as was done in [206,425–427]. This is sometimes reasonable, but always significantly more complicated. That is why the authors [421,422] chose the simplest way to screen the long-range stress field of the moving (first) disclination using the simplest self-screened disclination configuration (disclination dipole) among others which are observed in real materials.

Third, it has been supposed that the disclination dipole consists of mobile (first) and immobile (second) disclinations (Fig. 4.16). As with the first disclination, the second disclination may also represent the discontinuity of misorientation across the grain boundary, or it may be a triple junction disclination, etc. It may also be as mobile as the first disclination. This would not change the results of the model, because the authors [421, 422] have analysed only the energetic possibility for an elemental displacement of the first disclination. Both the disclinations could hardly be expected to make such elemental 'jumps' simultaneously. Therefore, one can treat one of them as mobile and the other as immobile.

Following [421, 422], let us consider the energetic characteristics of the grain boundary disclination motion under discussion (Fig. 4.16). The dipole of the grain boundary disclinations can be treated as a defect terminating a grain boundary dislocation wall of finite extent (Fig. 4.17), in which case an elementary transfer of the moving disclination occurs via the splitting of one grain boundary dislocation belonging to the wall into two lattice dislocations. The motion of the grain boundary disclination occurs under the action of an external mechanical load which, in the framework of the model, causes uniaxial stress parallel with the grain boundary plane and promotes

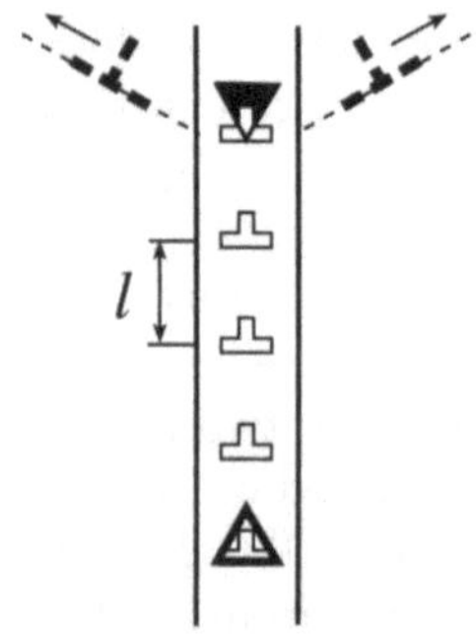

Fig. 4.17. Disclinations composing a dipole terminate a ragged wall of periodically spaced grain boundary dislocations (with period l) with either a crystal-lattice or DSC-lattice Burgers vector in small- and large-angle boundaries, respectively. An elementary transfer of a moving disclination, shown in Fig. 4.16, is accompanied by the splitting of one of the grain boundary dislocations into two lattice dislocations

the motion of the emitted dislocations (Fig. 4.16). The elementary transfer of a grain boundary disclination through a distance l, accompanied by emission of two lattice dislocations, is energetically favorable if the energy (per unit disclination length) W_2 of the defect configuration resulting from the transfer is lower than the energy W_1 of the pre-existing configuration (before the transfer): $\Delta W = W_2 - W_1 < 0$.

The pre-existing configuration represents a dipole of grain boundary disclinations separated by a distance L (the dipole arm). The energy W_1 of this system is given by

$$W_1 = E_{\mathrm{d}} = \frac{D\omega^2 L^2}{2}\left(\ln\frac{R}{L} + \frac{1}{2}\right) + NE_b^{\mathrm{c}}\,, \tag{4.17}$$

where the first term is the strain energy of the disclination dipole [see (4.1) in Sect. 4.1.3.1] and the second term is the summed energy of the cores of the N grain boundary dislocations which compose the ragged wall (Fig. 4.17). The core energy is approximated as $E_b^{\mathrm{c}} \approx Db^2/2$ [349].

The energy density W_2 of the dipole configuration resulting from the elementary transfer (Fig. 4.16) can be written as [321, 322]

$$W_2 = E'_{\mathrm{d}} + E_{b_1} + E_{b_2} + E_{\mathrm{d}}^{b_1} + E_{\mathrm{d}}^{b_2} + E_{b_2}^{b_1} + E_\sigma^{b_1} + E_\sigma^{b_2}\,. \tag{4.18}$$

Here E'_{d} denotes the strain energy of the resultant disclination dipole characterized by the dipole arm $L_1 = L - l$, E_{b_1} (E_{b_2}) the self-energy of the first (second) emitted dislocation, $E_{\mathrm{d}}^{b_1}$ ($E_{\mathrm{d}}^{b_2}$) the energy that characterizes the elastic interaction between the first (second) emitted dislocation and the disclination dipole, $E_{b_2}^{b_1}$ the energy that characterizes elastic interaction between the emitted dislocations, and $E_\sigma^{b_1}$ ($E_\sigma^{b_2}$) the work done by the external stress σ to transfer the first (second) dislocation to its position shown in Fig. 4.16.

The strain energy E'_{d} of the disclination dipole with arm L_1 is given by (4.17), substituting L for L_1 and N for $(N-1)$. The self-energies of the dislocations are given by (3.41) in Sect. 3.2.1.1, where the dislocation core radius is assumed to be the same for both the dislocations under consideration.

The other terms in the sum (4.18) are calculated in [422]. The final expressions for the energies $E_{\mathrm{d}}^{b_i}$, $E_{b_2}^{b_1}$, and $E_{\sigma}^{b_i}$ are

$$E_{\mathrm{d}}^{b_i} = \frac{D}{2}\omega b_i \cos\phi_i \left(L \ln \frac{R^2 + L^2 + 2RL\sin\phi_i}{p^2 + L^2 + 2pL\sin\phi_i} - l \ln \frac{R^2 + l^2 + 2Rl\sin\phi_i}{p^2 + l^2 + 2pl\sin\phi_i} \right) , \tag{4.19}$$

$$E_{b_2}^{b_1} = \frac{D}{2} b_1 b_2 \left\{ 1 + \cos(\phi_1 + \phi_2) \ln \frac{p^2 + R^2 + 2pR\cos(\phi_1 + \phi_2)}{2\mathrm{e}p^2\left[1 + \cos(\phi_1 + \phi_2)\right]} - \frac{2pR\sin^2(\phi_1 + \phi_2)}{p^2 + R^2 + 2pR\cos(\phi_1 + \phi_2)} \right\} , \tag{4.20}$$

$$E_{\sigma}^{b_i} = -\frac{\sigma}{2} b_i p \sin(2\phi_i) , \tag{4.21}$$

where e is the base of the natural logarithm and σ is the external normal stress (Fig. 4.16). Using (3.41) and (4.17)–(4.21), the authors [421, 422] found the characteristic energy difference ΔW:

$$\begin{aligned} \Delta W = \frac{D}{2} \Bigg\{ & \left(b_1^2 + b_2^2\right)\left(\ln \frac{R}{r_0} + 1\right) - b^2 - \frac{\sigma}{D} p\left[b_1 \sin(2\phi_1) + b_2 \sin(2\phi_2)\right] \\ & + b_1 b_2 \left[1 + \cos(\phi_1 + \phi_2) \ln \frac{p^2 + R^2 + 2pR\cos(\phi_1 + \phi_2)}{2\mathrm{e}p^2\left[1 + \cos(\phi_1 + \phi_2)\right]} - \frac{2pR\sin^2(\phi_1 + \phi_2)}{p^2 + R^2 + 2pR\cos(\phi_1 + \phi_2)} \right] \\ & + \omega^2 L_1^2 \left[\ln \frac{R}{L_1} + \frac{1}{2} - \frac{L^2}{L_1^2}\left(\ln \frac{R}{L} + \frac{1}{2}\right) \right] \\ & + \omega \sum_{i=1}^{2} b_i \cos\phi_i \left(L \ln \frac{R^2 + L^2 + 2RL\sin\phi_i}{p^2 + L^2 + 2pL\sin\phi_i} - l \ln \frac{R^2 + l^2 + 2Rl\sin\phi_i}{p^2 + l^2 + 2pl\sin\phi_i} \right) \Bigg\} . \end{aligned} \tag{4.22}$$

Equation (4.22) has been used to carry out numerical investigations of the dependence of ΔW on the parameters p, ϕ_1, ϕ_2, σ, R, and L for the system under consideration. The dependences $\Delta W(p)$ are shown in Fig. 4.18 for various values of the characteristic angles ϕ_1 and ϕ_2 related to the crystallography of the adjacent grains [422]. These dependences indicate that the angles ϕ_1 and ϕ_2 crucially influence the elementary transfer of a grain boundary disclination. In fact, the disclination transfer is energetically facilitated at

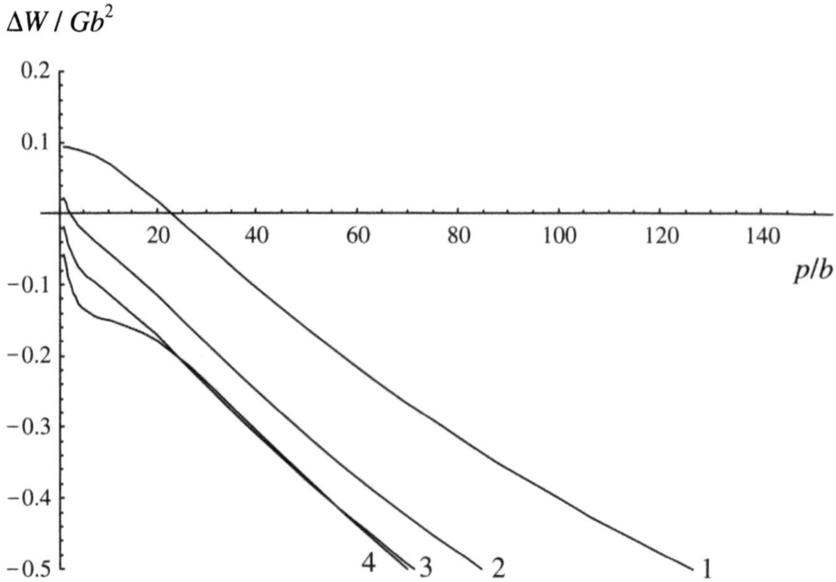

Fig. 4.18. Dependence of ΔW on the distance p moved by each of the emitted lattice dislocations for $L = 30b$, $\sigma = 10^{-3}G$, $R = 10^5 b$, $\omega = 0.1$, and the following values of the characteristic angles: $\phi_1 = \phi_2 = 45°$ (curve 1), $\phi_1 = 30°$ and $\phi_2 = 40°$ (curve 2), $\phi_1 = 20°$ and $\phi_2 = 30°$ (curve 3), and $\phi_1 = \phi_2 = 2°$ (curve 4)

low angles and hampered with rising ϕ_1 and ϕ_2. At large values of ϕ_1 and ϕ_2 (roughly $> 50°$), there exists an energetic barrier for motion of the emitted lattice dislocations, and the disclination transfer is energetically unfavorable. In the range of ϕ_1 and ϕ_2 from 0° to roughly 20°, the dislocation motion has a barrierless character with $\Delta W (< 0)$ decreasing with distance p moved along the dislocation path (Fig. 4.18). In the range of ϕ_1 and ϕ_2 from roughly 20° to 50°, the disclination transfer is either favorable or unfavorable, depending on other parameters (L, ω, σ) of the system. In this situation, the characteristic energy difference $\Delta W(p = b)$ at the starting point of the dislocation motion is highly sensitive to both the distance L between the disclinations and their strength ω (see Figs. 4.19 and 4.20) [421, 422].

At the same time, $\Delta W(p = b)$ depends weakly on the external stress σ (see Fig. 4.21) [422]. The influence of the stress σ on the characteristic energy difference $\Delta W(p)$ increases in moving along the dislocation path (increasing p) at some intermediate values (close to 45°) of angles ϕ_1 and ϕ_2. The energy difference $\Delta W(p)$ at low values of angles ϕ_1 and ϕ_2 is weakly sensitive to σ, as shown in Fig. 4.22 [422]. This follows naturally from the geometry of the system (Fig. 4.16).

Consequently, the probability of a grain boundary dislocation splitting into a pair of lattice dislocations, together with their emission and corresponding motion of the grain boundary disclination, increases with decreasing values of angles ϕ_1 and ϕ_2, and with growth of the dipole arm L and

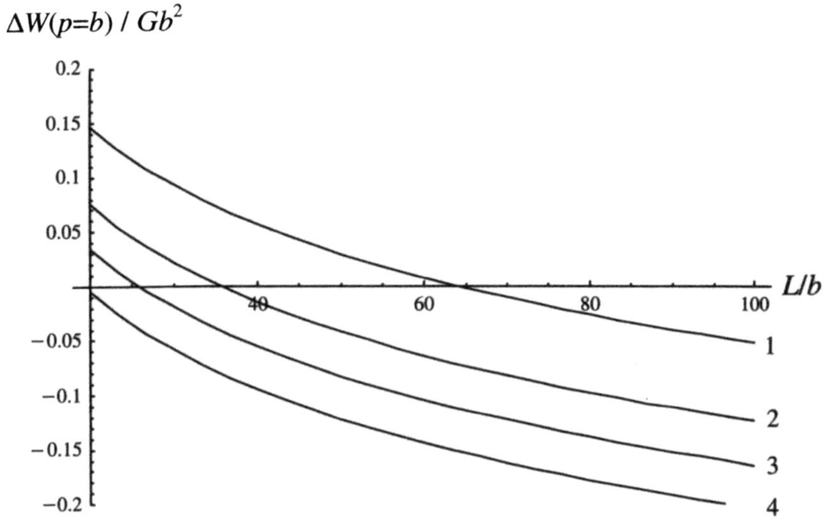

Fig. 4.19. Dependence of ΔW on the distance L between grain boundary disclinations at the initial stage of the dislocation emission (at $p = b$), for $\sigma = 10^{-3}G$, $R = 10^5 b$, $\omega = 0.1$, and the following values of the characteristic angles: $\phi_1 = \phi_2 = 45°$ (curve 1), $\phi_1 = 30°$ and $\phi_2 = 40°$ (curve 2), $\phi_1 = 20°$ and $\phi_2 = 30°$ (curve 3), and $\phi_1 = \phi_2 = 2°$ (curve 4)

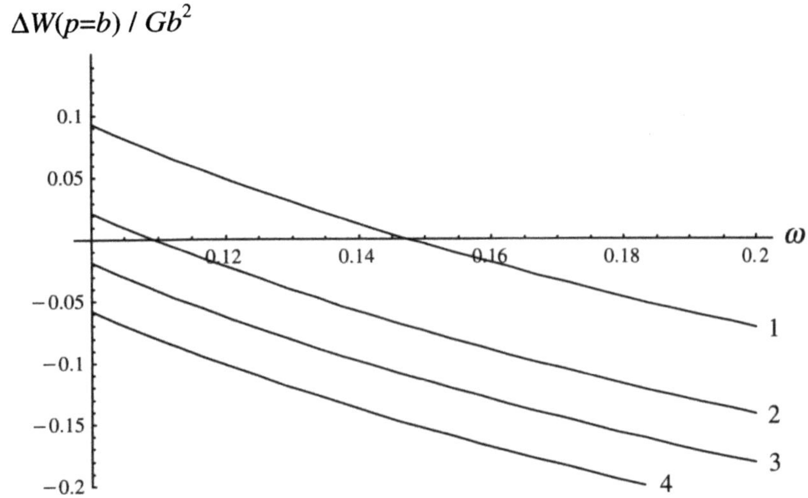

Fig. 4.20. Dependence of ΔW on the disclination strength ω at the initial stage of the dislocation emission (at $p = b$), for $\sigma = 10^{-3}G$, $R = 10^5 b$, $L = 30b$, and the following values of the characteristic angles: $\phi_1 = \phi_2 = 45°$ (curve 1), $\phi_1 = 30°$ and $\phi_2 = 40°$ (curve 2), $\phi_1 = 20°$ and $\phi_2 = 30°$ (curve 3), and $\phi_1 = \phi_2 = 2°$ (curve 4)

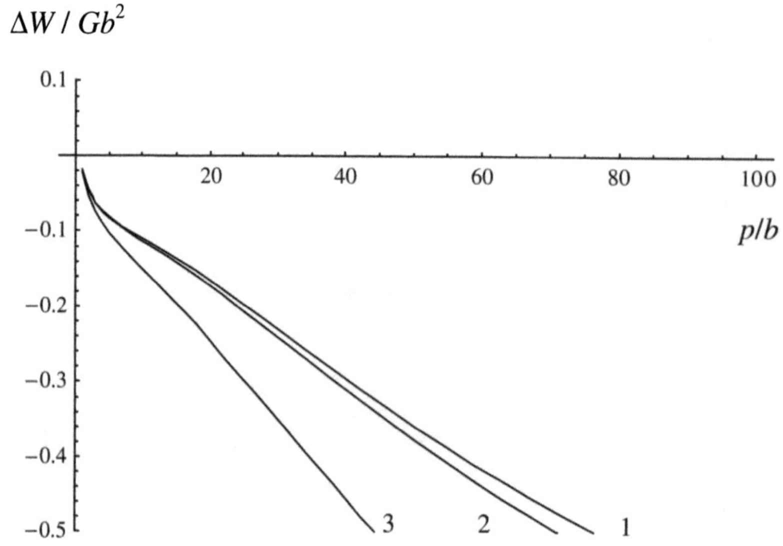

Fig. 4.21. Dependence of ΔW on the distance p moved by the emitted dislocations for $\phi_1 = 30°$ and $\phi_2 = 40°$, $R = 10^5\,b$, $L = 30b$, $\omega = 0.1$, and the following values of the applied stress $\sigma/G = 10^{-4}$, 10^{-3} and 10^{-2} (curves 1, 2 and 3, respectively)

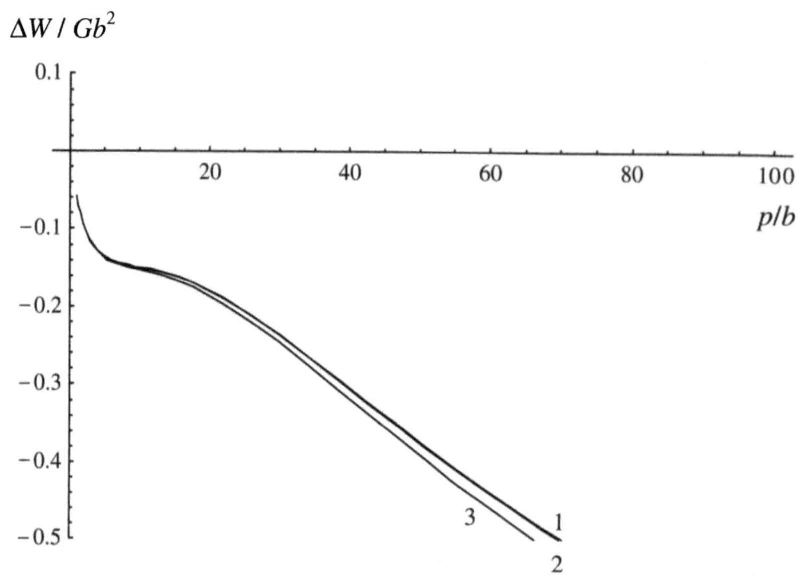

Fig. 4.22. Dependence of ΔW on the distance p moved by the emitted dislocations, for $\phi_1 = \phi_2 = 2°$, $R = 10^5\,b$, $L = 30b$, $\omega = 0.1$, and the following values of the applied stress $\sigma/G = 10^{-4}$, 10^{-3} and 10^{-2} (curves 1, 2 and 3, respectively)

disclination strength ω. The grain boundary disclination motion has been proved to be an energetically favorable process over rather wide ranges of parameters that characterize the defect configuration under consideration. In contrast to the previously considered situation where disclination motion was associated with (ordered in space and time) absorption of dislocations by grain boundaries [56, 263, 409], the model of disclination motion associated with emission of dislocation pairs [421, 422] does not require any correlated flux of dislocations from grain interiors to grain boundaries. The suggested micromechanism for the disclination motion (Fig. 4.16) can be responsible for experimentally observed rotations of grains in fine-grained materials under (super)plastic deformation and thermal treatment.

4.2.2 Motion of a Dipole of Grain Boundary Disclinations

The model of motion of a grain boundary disclination [421,422] discussed in the previous section is easily extended to the case of conservative motion of a dipole of such disclinations under the action of external loading [419, 420]. The motion of dipoles of grain boundary disclinations may be considered as similar to the motion of (super)dislocations and essentially contributes to high-strain deformation (Fig. 4.23) [56, 263]. Experimental evidence for the existence of disclination dipoles in typical NCMs has recently been demonstrated in direct atomic-level HREM observations by Murayama et al. [218].

Following the model [419, 420], one can assume that the rotational deformation in a fine-grained material occurs via the stress-induced motion of a dipole of grain boundary disclinations. In the framework of the model, the dipole consists of two grain boundary disclinations with strengths $+\omega$

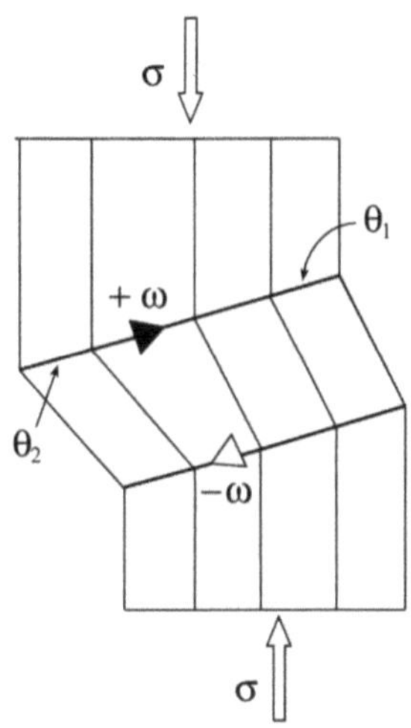

Fig. 4.23. The motion of a disclination dipole may be considered as similar to the motion of a (super)dislocation

and $-\omega$, respectively, and has arm L. The disclination of strength $+\omega$ ($+\omega$-disclination) transfers through the distance l due to the emission of two lattice dislocations with Burgers vectors $\boldsymbol{b}$ and $\boldsymbol{b}_1$ into the adjacent grains (Fig. 4.24) [419, 420]. The dislocation with Burgers vector $\boldsymbol{b}$ moves along its gliding plane towards the disclination of strength $-\omega$ ($-\omega$-disclination). This disclination moves through the distance l in the same direction as the $+\omega$-disclination due to the emission of two lattice dislocations with Burgers vectors $-\boldsymbol{b}$ and $\boldsymbol{b}_2$ into the abutting grains. The dislocations with Burgers vectors $\boldsymbol{b}$ and $-\boldsymbol{b}$ emitted by the $+\omega$- and $-\omega$-disclinations, respectively, meet each other and annihilate.

Of course, this model is very idealistic, even more than the one considered in Sect. 4.2.1. In a real NCM fabricated by a severe deformation method, the grain boundary disclinations with various values of strength are distributed

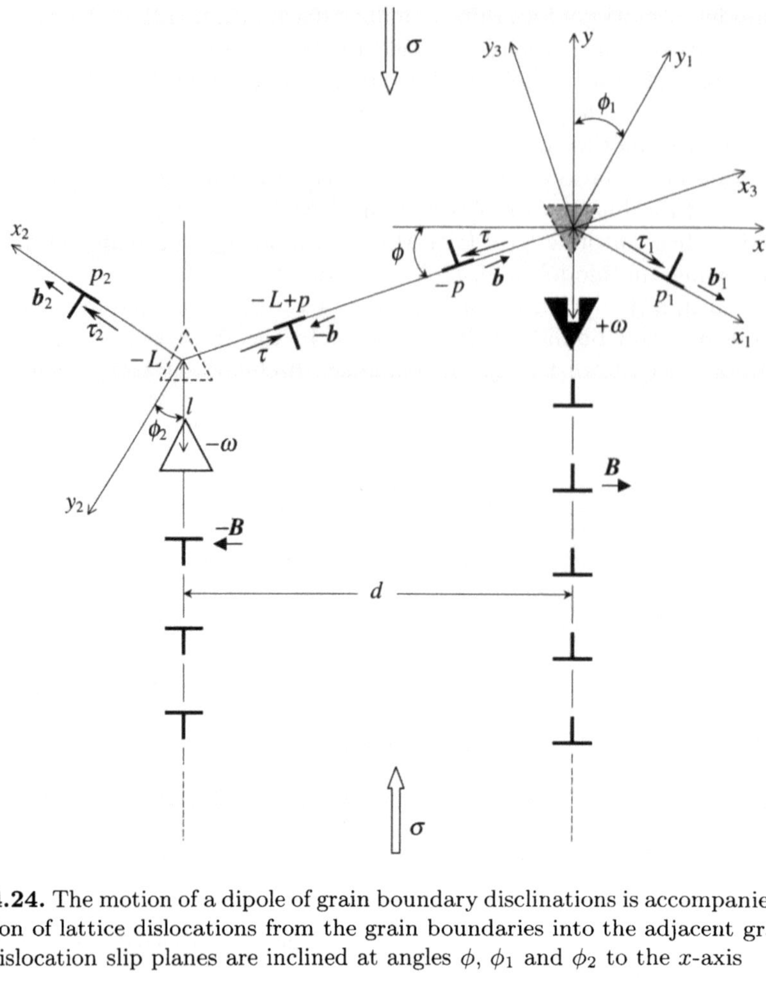

Fig. 4.24. The motion of a dipole of grain boundary disclinations is accompanied by emission of lattice dislocations from the grain boundaries into the adjacent grains. The dislocation slip planes are inclined at angles ϕ, ϕ_1 and ϕ_2 to the x-axis

in a rather disordered manner along non-equilibrium grain boundaries and the emitted lattice dislocations do not necessarily annihilate. In addition, grain interiors of a fine-grained material contain many lattice dislocations generated during severe plastic deformation. These dislocations are capable of interacting and annihilating with the dislocations emitted by grain boundary disclinations. Thus, real processes occurring during motion of grain boundary disclinations can be very complicated and different from those described by the model [419, 420]. However, this simplified model is convenient for a strict mathematical analysis and effective for understanding the key peculiarities of grain boundary disclination motion accompanied by emission of lattice dislocations. Moreover, it can serve as a basis for further detailed study of the evolution of grain boundary defects in plastically deformed NCMs.

The cooperative motion of the grain boundary disclinations composing the dipole causes the plastic deformation of an NCM. The dipole motion is energetically favorable if the difference between the system energies (per unit disclination length) after (W_2) and before (W_1) the dipole elementary transfer is negative: $\Delta W = W_2 - W_1 < 0$. The corresponding calculation of ΔW is very similar to that described in Sect. 4.2.1, so below we give only a few intermediate formulas and the final result. The full calculation procedure can be found in [419].

The energy W_1 of the system in its initial state (before the dipole transfer) is the strain energy of the disclination dipole, given by the first term in (4.17). The disclination dipole motion is accompanied by the emission of lattice dislocations, in which case the energy density W_2 of the defect system comprises the five terms [419, 420]

$$W_2 = E_{\mathrm{d}} + E_{b_1} + E_{b_2} + E_{bb} + E_{\mathrm{int}} \,. \tag{4.23}$$

Here $E_{\mathrm{d}} = W_1$ is the strain energy of the disclination dipole, which does not change during its correlated motion, E_{b_1} and E_{b_2} are the self-energies of the emitted dislocations with Burgers vectors $\boldsymbol{b}_1$ and $\boldsymbol{b}_2$, respectively, E_{bb} is the self-energy of the dislocation dipole consisting of the emitted dislocations with Burgers vectors $\boldsymbol{b}$ and $-\boldsymbol{b}$, and E_{int} is the summed energy characterizing all interactions between the defects composing the system (Fig. 4.24) (except between the two disclinations whose interaction energy is included in E_{d}, and between the $\boldsymbol{b}$- and $-\boldsymbol{b}$-dislocations whose interaction energy is included in E_{bb}), and between the defects and the applied stress σ. The last component can be decomposed into nine terms as follows [419]:

$$E_{\mathrm{int}} = E_{\mathrm{d}}^{b_1} + E_{\mathrm{d}}^{b_2} + E_{\mathrm{d}}^{bb} + E_{b_2}^{b_1} + E_{bb}^{b_1} + E_{bb}^{b_2} + E_{\sigma}^{b_1} + E_{\sigma}^{b_2} + E_{\sigma}^{bb} \,. \tag{4.24}$$

Here $E_{\mathrm{d}}^{b_1}$, $E_{\mathrm{d}}^{b_2}$ and E_{d}^{bb} denote the energies characterizing the interaction of the disclination dipole with the $\boldsymbol{b}_1$- and $\boldsymbol{b}_2$-dislocations, and the dipole of $\pm\boldsymbol{b}$-dislocations, respectively. $E_{b_1}^{b_2}$ is the energy of the interaction between the $\boldsymbol{b}_1$- and $\boldsymbol{b}_2$-dislocations. $E_{bb}^{b_1}$ and $E_{bb}^{b_2}$ denote the energies that characterize the interaction of the dislocation dipole with these dislocations. $E_\sigma^{b_1}$, $E_\sigma^{b_2}$ and E_σ^{bb} are the energies characterizing the interaction of the applied stress σ with the emitted $\boldsymbol{b}_1$- and $\boldsymbol{b}_2$-dislocations, and the dislocation dipole, respectively.

Consider the energies featuring on the right-hand side of (4.23), taking (4.24) into account. The self-energies of the $\boldsymbol{b}_1$- and $\boldsymbol{b}_2$-dislocations are given by (3.41). The self-energy of the dipole of $\pm\boldsymbol{b}$-dislocations is calculated as the work done to generate these dislocations in their summed elastic field, which eventually gives [419]

$$E_{bb} = Db^2\left(\ln\frac{|L-2p-r_c|}{r_c}+1\right), \tag{4.25}$$

where the last term 1 is added to take the contributions of the dislocation cores into account.

The energies of interaction between the disclination dipole and the lattice dislocations are calculated in a similar way, as the work done to generate these dislocations in the stress field of the disclination dipole. In doing so, one has to use the different coordinate systems shown in Fig. 4.24. These rather cumbersome calculations result in [419]

$$E_{\mathrm{d}}^{b_1} = \frac{D\omega b_1}{2}\Psi(p_1,\phi,\phi_1), \tag{4.26}$$

$$E_{\mathrm{d}}^{b_2} = \frac{D\omega b_2}{2}\Psi(p_2,-\phi,-\phi_2), \tag{4.27}$$

$$E_{\mathrm{d}}^{bb} = -\frac{D\omega bl}{2}\cos\phi\ln\frac{(L-p)^4+l^4+2(L-p)^2l^2\cos(2\phi)}{p^4+l^4+2p^2l^2\cos(2\phi)}. \tag{4.28}$$

introducing the function

$$\begin{aligned}\Psi(p_1,\phi,\phi_1) = {}& \left[l\cos\phi_1 + L\sin(\phi+\phi_1)\right] \\ & \times\ln\frac{L^2+R^2+l^2+2LR\cos(\phi+\phi_1)+2Ll\sin\phi-2Rl\sin\phi_1}{L^2+p_1^2+l^2+2Lp_1\cos(\phi+\phi_1)+2Ll\sin\phi-2p_1l\sin\phi_1} \\ & -l\cos\phi_1\ln\frac{R^2+l^2-2Rl\sin\phi_1}{p_1^2+l^2-2p_1l\sin\phi_1}.\end{aligned} \tag{4.29}$$

Here l denotes the distance between the grain boundary dislocations which compose the ragged dislocation walls and the disclinations at the edges (Fig. 4.24). This distance may be written in terms of the magnitude of the Burgers vector $\boldsymbol{B}$ of the grain boundary dislocations and the disclination strength ω as

$$l \approx B/\omega. \tag{4.30}$$

In its turn, the Burgers vector $\boldsymbol{B}$ is connected with other geometric parameters of this defect configuration by the relations

$$B = b_1\cos\phi_1 + b\cos\phi \quad \text{and} \quad \frac{b_1}{b} = \frac{\sin\phi}{\sin\phi_1}.$$

Substituting them into (4.30) gives

$$l \approx \frac{b\sin(\phi + \phi_1)}{\omega \sin\phi_1} \,. \tag{4.31}$$

The energy of interaction $E^{b_1}_{b_2}$ between the $\boldsymbol{b}_1$- and $\boldsymbol{b}_2$-dislocations is also calculated as the work done to generate one dislocation in the stress field of the other. The result is [419]

$$E^{b_1}_{b_2} = -Db_1b_2\left[\frac{S_1S_2}{P^2} + \cos(\phi_1 - \phi_2)\ln\frac{R}{|P|}\right] , \tag{4.32}$$

using the notation

$$\begin{gathered} S_1 = p_1\sin(\phi_1 - \phi_2) - L\sin(\phi + \phi_2) \,, \\ S_2 = p_2\sin(\phi_1 - \phi_2) + L\sin(\phi + \phi_1) \,, \\ P^2 = L^2 + p_1^2 + p_2^2 + 2p_1p_2\cos(\phi_1 - \phi_2) + 2Lp_1\cos(\phi + \phi_1) + 2Lp_2\cos(\phi + \phi_2) \,. \end{gathered} \tag{4.33}$$

The energy of interaction $E^{b_i}_{bb}$ of the $\boldsymbol{b}_i$-dislocation with the dipole of $\pm\boldsymbol{b}$-dislocations is determined in a similar way, giving [419]

$$E^{b_i}_{bb} = \frac{Dbb_i}{2}\Phi(p_i, \phi, \phi_i) \,, \tag{4.34}$$

where $i = 1, 2$ and

$$\begin{aligned} \Phi(p_i, \phi, \phi_i) &= \cos(\phi + \phi_i)\ln\frac{(L-p)^2 + p_i^2 + 2(L-p)p_i\cos(\phi + \phi_i)}{p^2 + p_i^2 + 2pp_i\cos(\phi + \phi_i)} \\ &- \frac{2p_i(L-2p)(p^2 + p_i^2 - Lp)\sin^2(\phi + \phi_i)}{\left[(L-p)^2 + p_i^2 + 2(L-p)p_i\cos(\phi + \phi_i)\right]\left[p^2 + p_i^2 + 2pp_i\cos(\phi + \phi_i)\right]} \,. \end{aligned} \tag{4.35}$$

The energies of interaction of the lattice dislocations with the external stress σ are given by the simple formulas

$$E^{b_i}_{\sigma} = -\frac{\sigma}{2}b_ip_i\sin(2\phi_i) \,, \tag{4.36}$$

$$E^{b}_{\sigma} = -\frac{\sigma}{2}bp\sin(2\phi) \,. \tag{4.37}$$

All the terms in the sums (4.23) and (4.24) are thus obtained. The energy difference $\Delta W = W_2 - W_1$ may then be written in the following final form [419]:

$$\Delta W = \frac{D}{2}\Bigg\{(b_1^2 + b_2^2)\left(\ln\frac{R}{r_c} + 1\right) + 2b^2\left(\ln\frac{|L-2p-r_c|}{r_c} + 1\right)$$
$$-2b_1b_2\left[\frac{S_1S_2}{P^2} + \cos(\phi_1 - \phi_2)\ln\frac{R}{|P|}\right]$$
$$-\frac{\sigma}{D}\big[2bp\sin(2\phi) + b_1p_1\sin(2\phi_1) + b_2p_2\sin(2\phi_2)\big]$$
$$+b\big[b_1\Phi(p_1,\phi,\phi_1) + b_2\Phi(p_2,\phi,\phi_2)\big]$$
$$+\omega\big[b_1\Psi(p_1,\phi,\phi_1) + b_2\Psi(p_2,-\phi,-\phi_2)\big]$$
$$-\,\omega bl\cos\phi\ln\frac{(L-p)^4 + l^4 + 2(L-p)^2l^2\cos(2\phi)}{p^4 + l^4 + 2p^2l^2\cos(2\phi)}\Bigg\}\,. \quad (4.38)$$

The results of numerical calculations carried out using (4.38) are presented in Figs. 4.25 and 4.26 [419,420]. The dependences of the characteristic angles ϕ_1 and ϕ_2 on the angle ϕ are shown in Fig. 4.25, in the case where $\phi_1 = \phi_2$. These dependences, for the given values of ϕ, indicate the angles ϕ_1 and ϕ_2 at which the disclination dipole motion (Fig. 4.24) is characterized by the lowest value $\Delta W_{\min}$ (where $\Delta W < 0$) of the energy difference (4.38) that corresponds to the start of the disclination dipole motion. (In other words, $\Delta W_{\min}$ is the largest energy gain due to the start of motion of emitted lattice

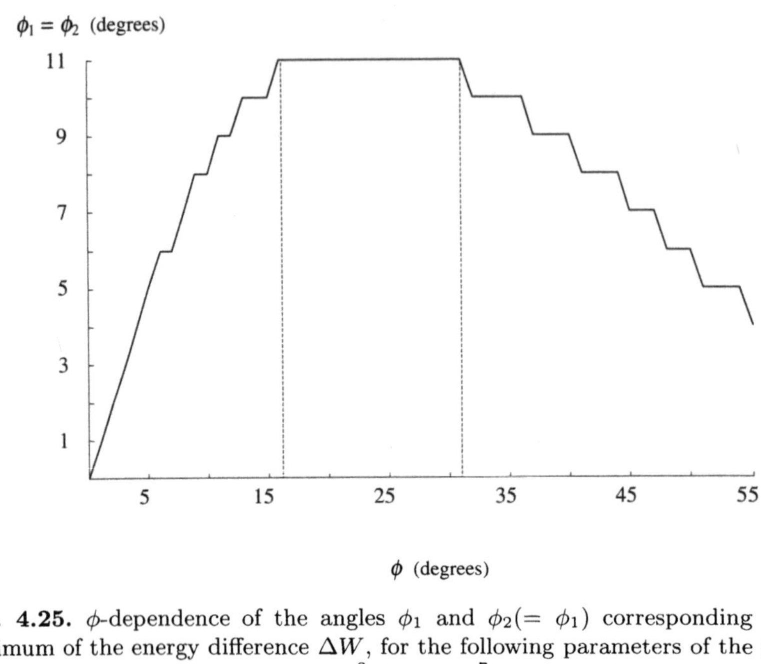

Fig. 4.25. ϕ-dependence of the angles ϕ_1 and $\phi_2 (= \phi_1)$ corresponding to the minimum of the energy difference ΔW, for the following parameters of the model: $\omega = 0.1$, $p_1 = p_2 = p = b$, $\sigma = 5 \times 10^{-3}G$, $R = 10^7 b$ and $L = 1000b$

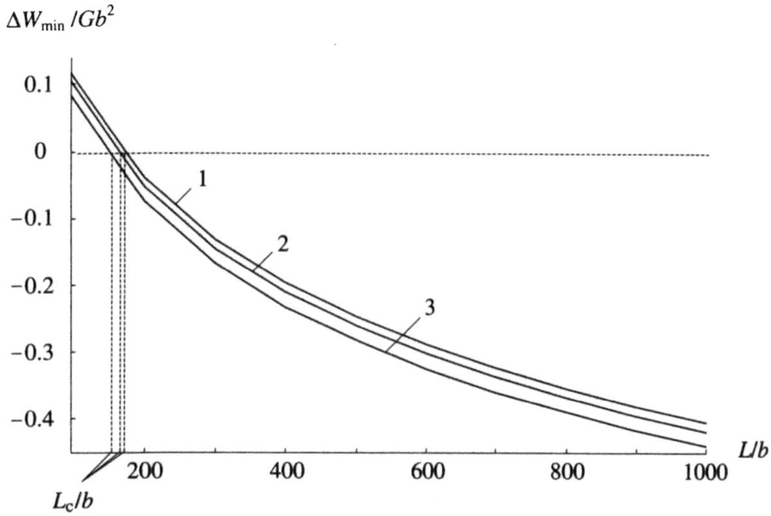

Fig. 4.26. Dependence of the energy difference $\Delta W_{\min}$ on the disclination dipole arm L for the following parameters of the model: $\omega = 0.1$, $p_1 = p_2 = p = b$, $\sigma = 5 \times 10^{-3}G$, $R = 10^7 b$ and $\phi = 30°$ (curve 1), 40° (curve 2) and 50° (curve 3)

dislocations, associated with the process in question.) As can be seen from Fig. 4.25, these angles ϕ_1 and $\phi_2(=\phi_1)$ increase with rising ϕ from 0° to $\approx 15°$. They then remain constant at $\approx 11°$ in the range of ϕ from $\approx 15°$ to $\approx 30°$, and decrease from $\approx 11°$ to $\approx 4°$ with rising ϕ from $\approx 30°$ to $\approx 55°$.

The angles ϕ_1 and $\phi_2(=\phi_1)$ which correspond to the minimum of the energy difference ΔW depend weakly on the other parameters L, σ, and ω of the system. At the same time, the minimum value $\Delta W_{\min}$ of this characteristic energy difference is sensitive to the parameters discussed. In Fig. 4.26, the dependences of $\Delta W_{\min}$ on the disclination dipole arm L are presented for various values of the angle ϕ [420]. As can be seen from Fig. 4.25, the disclination dipole motion is energetically favorable ($\Delta W_{\min} < 0$) if the dipole arm L exceeds a critical value L_c, which is ϕ-dependent and close to about $150b \approx 50$ nm. In the range of ϕ from 1° to roughly 30°, the dependences $\Delta W_{\min}(L)$ coincide. They are shown as curve 1 in Fig. 4.26. The dependences $\Delta W_{\min}(L)$ at $\phi = 40°$ and 50° (curves 2 and 3, respectively, in Fig. 4.26) are different from those at $\phi \leq 30°$. They indicate that $\Delta W_{\min}(L)$ decreases slightly as ϕ rises from 30° to 50°. To summarize, $\Delta W_{\min} < 0$, i.e., the disclination dipole motion (Fig. 4.24) is an energetically favorable process over a rather wide range of values of the parameters characterizing the defect configuration under consideration.

In [419, 420], it has been shown theoretically that the rotational mode of plastic flow in the fine-grained materials produced by severe plastic deformation can effectively occur via grain boundary disclination dipole motion associated with the emission of lattice dislocation pairs into the adjacent grains (Fig. 4.24). In contrast to the previously considered [56,263,409] situation where the disclination motion is associated with the absorption (ordered in space and time) of lattice dislocations by grain boundaries, the disclination dipole motion associated with the emission of lattice dislocation pairs does not require any correlated flux of dislocations from grain interiors to grain boundaries. This micromechanism for the disclination motion can be responsible for the rotations of grains observed experimentally [189,196,332,409,413–416] in fine-grained materials under (super)plastic deformation and thermal treatment.

4.2.3 Crossover from Grain Boundary Sliding to Rotational Deformation

In the previous sections we have considered the motion of partial disclinations which is realized by absorption (Sect. 4.1) or emission (Sects. 4.2.1 and 4.2.2) of lattice dislocations. Thus, cooperative action of the translational (dislocation gliding) and rotational (disclination motion) modes of plastic deformation in NCMs has been assumed. However, as we have already mentioned above, with grain refinement below a critical grain size d^* the gliding of lattice dislocations ceases to play any significant role in NCMs and new deformation micromechanisms such as grain boundary sliding and diffusion mass transfer are activated. The question arises as to how the rotational mode of plasticity can develop in these conditions. In the present section we discuss the results of recent theoretical investigations into the relation between the grain boundary sliding and rotational deformation which occurs through generation and motion of partial grain boundary disclinations by means of the climb of grain boundary dislocations along grain boundaries in NCMs [428]. It is demonstrated that, under special conditions of plastic deformation, the crossover from the translational micromechanism of plasticity (grain boundary sliding) to the rotational one (grain rotation) is energetically favorable.

4.2.3.1 Splitting of Gliding Grain Boundary Dislocations into Climbing Dislocations at Triple Junctions. (Small-Scale View)

Grain boundary sliding, treated as the dominant mode of superplasticity in nano- and microcrystalline materials, occurs via motion of gliding grain boundary dislocations (see Sect. 3.2). They have Burgers vectors that are parallel to the corresponding grain boundary planes along which these dislocations glide. Triple junctions of grain boundaries, where grain boundary planes change their orientations, serve as obstacles for the grain boundary dislocation motion. In these circumstances, splitting of gliding grain boundary dislocations can occur at triple junctions, resulting in the formation of sessile dislocations and gliding dislocations providing further grain boundary sliding along the adjacent grain boundaries (see Sects. 2.2.4 and 3.2). However, in general, grain boundary dislocations stopped at a triple junction are also capable of splitting into climbing grain boundary dislocations (Fig. 4.27) [428].

When this process occurs repeatedly at a triple junction, it results in the formation of two walls of grain boundary dislocations climbing along the grain boundaries adjacent to the triple junction. The climbing grain boundary dislocation walls cause the rotational deformation, in which case the repeated

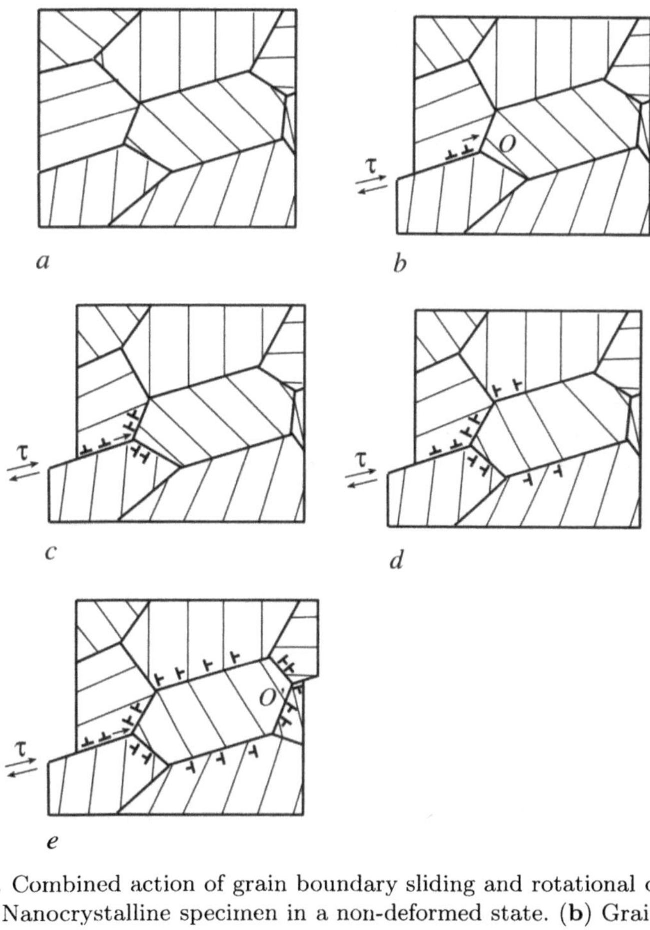

Fig. 4.27. Combined action of grain boundary sliding and rotational deformation mode. (**a**) Nanocrystalline specimen in a non-deformed state. (**b**) Grain boundary sliding occurs via motion of gliding grain boundary dislocations under shear stress action. (**c**) Gliding dislocations split into climbing dislocations at triple junction O of grain boundaries. (**d**) The splitting of gliding grain boundary dislocations occurs repeatedly, causing the formation of walls of grain boundary dislocations whose climb is accompanied by crystal lattice rotation in a grain. (**e**) Climbing dislocations reach triple junction O', where they converge into gliding dislocations causing further grain boundary sliding

splitting of gliding grain boundary dislocations at the triple junction provides the crossover from grain boundary sliding to the rotational deformation mode. This process can be spread over the grain, which has to rotate as a whole through an angle. Hence, the stopped grain boundary sliding can stimulate plastic rotation of the neighbouring grain. Obviously, this mechanism may only be effective under the condition of intensive grain boundary diffusion of vacancies, which must be able to provide the necessary velocity of grain boundary dislocation climb.

The reality of the model [428] may be estimated from the viewpoint of thermodynamics, by analysing the energetic favoritism of this process and obtaining only the necessary conditions for its realization, or kinetics, which allows one to find sufficient conditions. In the framework of the first approach, which has already been demonstrated many times throughout this book, the problem is solved relatively simply and gives new results (see below). The second approach involves stating and solving a very complicated problem of grain boundary diffusion for vacancies in a system with many sources and sinks, whilst accounting for the highly inhomogeneous field of elastic stresses of the dislocation–disclination structure that evolves in space and time. Such a problem has neither been stated nor solved yet. Using the first approach, the authors [428] have considered the very beginning of the transition from grain boundary sliding to the formation of the final disclination structure, when the 'head' grain boundary dislocation in a pile-up splits into a pair of new grain boundary dislocations (small-scale view, see Sect. 4.2.3.1), and the very end of this transition, when two pile-ups of grain boundary dislocations at opposite sides of a grain are totally transformed into a system of grain boundary disclination dipoles (large-scale view, see Sect. 4.2.3.2).

Following the model [428], consider a pile-up of n_c grain boundary dislocations with Burgers vectors $\boldsymbol{b}$ near a triple junction of grain boundaries under the action of an external shear stress τ (Fig. 4.28a). Assume that the head grain boundary dislocation splits into two new grain boundary dislocations with Burgers vectors $\boldsymbol{b}_1$ and $\boldsymbol{b}_2$ which can climb along the adjacent grain boundaries (Fig. 4.28b). The splitting process may be characterized by the difference $\Delta W = W_2 - W_1$ between the energies of the final (W_2, Fig. 4.28b) and initial (W_1, Fig. 4.28a) states of the defect configuration. The splitting is energetically favorable (unfavorable), if $\Delta W < 0$ ($\Delta W > 0$, respectively). The equation $\Delta W = 0$ gives a set of critical values of parameters for the defect configuration, at which the splitting becomes energetically favorable.

The total energy W_1 of the initial defect configuration (per unit dislocation length) consists of two terms [428]

$$W_1 = E_1^{\mathrm{pu}} + E_{\Sigma 1}^{b-b} , \tag{4.39}$$

where E_1^{pu} is the sum of the self-energies of n_c grain boundary dislocations composing the pile-up, and $E_{\Sigma 1}^{b-b}$ is the energy that characterizes pairwise interactions between all these dislocations.

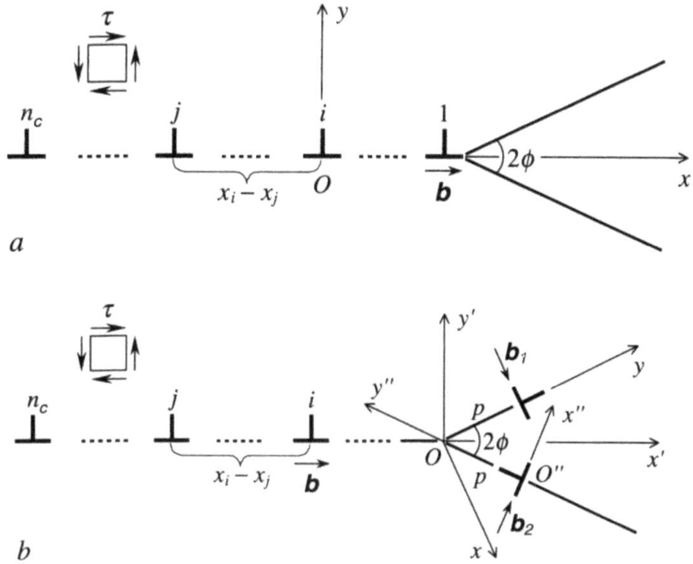

Fig. 4.28. Splitting of (**a**) the head dislocation of a grain boundary dislocation pile-up at a triple junction into (**b**) two dislocations that climb along the adjacent grain boundaries. The coordinate systems used in the calculations are Oxy, $Ox'y'$ and $O''x''y''$

The first term is written at once as [349]

$$E_1^{\mathrm{pu}} = n_{\mathrm{c}} \frac{Db^2}{2} \left(\ln \frac{R}{b} + 1 \right) . \tag{4.40}$$

In order to find the second term, let us first calculate the energy E_{ij}^{b-b} that characterizes the elastic interaction between the i th and j th grain boundary dislocations of the pile-up, with distance $(x_i - x_j)$ between them (Fig. 4.28a). The positions x_i and x_j of these dislocations may be calculated [429] with the help of the Laguerre polynomials $L_n(x)$, which are determined as solutions of the differential equation

$$x^2 \frac{\mathrm{d}^2 y}{\mathrm{d}x^2} + (1 - x) \frac{\mathrm{d}y}{\mathrm{d}x} + ny = 0 . \tag{4.41}$$

The first derivative of the Laguerre polynomial $L_n'(x)$ is a solution of the equation

$$x \frac{\mathrm{d}^2 y}{\mathrm{d}x^2} + (2 - x) \frac{\mathrm{d}y}{\mathrm{d}x} + (n - 1)y = 0 , \tag{4.42}$$

and is given by the formula

$$L_n'(x) = - \sum_{k=0}^{n-1} \frac{n!(-x)^k}{k!(k+1)!(n-k-1)!} . \tag{4.43}$$

Eshelby et al. [429] showed that the roots of the first derivative of the Laguerre polynomial $L'_n(x)$ determine the equilibrium positions of dislocations in a discrete dislocation pile-up. Thus, by calculating numerically the positions of the grain boundary dislocations in the pile-up for the given values of n_c and τ (Fig. 4.28a), one can also find the energy of elastic interaction between two grain boundary dislocations as the work done to transfer one dislocation from a free surface of a solid to its current position in the stress field created by the other dislocation. A similar approach has been used to analyze dislocation interactions within a grain boundary dislocation pile-up in [309]. In doing so, the energy E_{ij}^{b-b} of interaction between the i th and j th grain boundary dislocations has been found as [428]

$$E_{ij}^{b-b} = Db^2 \ln \frac{R}{x_i - x_j} . \tag{4.44}$$

The energy $E_{\Sigma 1}^{b-b}$ that characterizes pairwise interactions between all grain boundary dislocations belonging to the pile-up (Fig. 4.28a) is the sum of energies E_{ij}^{b-b} over the grain boundary dislocation indices i and j ($i \neq j$) [428]:

$$E_{\Sigma 1}^{b-b} = \sum_{i=1}^{n_c-1} \sum_{j=i+1}^{n_c} E_{ij}^{b-b} = Db^2 \sum_{i=1}^{n_c-1} \sum_{j=i+1}^{n_c} \ln \frac{R}{x_i - x_j} . \tag{4.45}$$

The energy of the defect configuration (Fig. 4.28b) resulting from the splitting of the head dislocation belonging to the grain boundary dislocation pile-up comprises nine terms [428]:

$$W_2 = E_2^{\mathrm{pu}} + E_{\Sigma 2}^{b-b} + E_{\mathrm{s}}^{b_1} + E_{\mathrm{s}}^{b_2} + E_{\Sigma 2}^{b-b_1} + E_{\Sigma 2}^{b-b_2} + E^{b_1-b_2} + E^{\tau-b_1} + E^{\tau-b_2} , \tag{4.46}$$

where E_2^{pu} is the sum of all self-energies of the grain boundary dislocations belonging to the pile-up after the splitting. $E_{\Sigma 2}^{b-b}$ is the energy of pairwise interactions between all grain boundary dislocations composing the pile-up. $E_{\mathrm{s}}^{b_1}$ and $E_{\mathrm{s}}^{b_2}$ are the self-energies of the two grain boundary dislocations resulting from the splitting of the head dislocation. $E_{\Sigma 2}^{b-b_1}$ and $E_{\Sigma 2}^{b-b_2}$ are the energies that characterize interaction between grain boundary dislocations of the pile-up and the grain boundary dislocations resulting from the splitting and characterized by Burgers vectors $\boldsymbol{b}_1$ and $\boldsymbol{b}_2$, respectively. $E^{b_1-b_2}$ is the energy of interaction between these latter (climbing) dislocations, and $E^{\tau-b_1}$ and $E^{\tau-b_2}$ are the effective work done by the shear stress τ to transfer these two grain boundary dislocations along the grain boundaries adjacent to the triple junction.

The pile-up after the splitting of its head dislocation (Fig. 4.28b) contains $(n_c - 1)$ grain boundary dislocations. To simplify the calculation of the energy characteristics of these dislocations, the authors [428] assumed that the dislocation positions remain unchanged during the splitting of the head dislocation. This assumption is reasonable at least in the situation where the grain boundary dislocations resulting from the splitting are close to the triple junction (Fig. 4.28b). Then the first term of the sum (4.46), E_2^{pu}, is directly obtained from (4.40) by replacing n_{c} by $(n_{\mathrm{c}}-1)$. The second term, $E_{\Sigma 2}^{b-b}$, follows from (4.45) by substituting $i=1$ for $i=2$ in the lower limit of the first sum.

The self-energies of the new climbing grain boundary dislocations, $E_{\mathrm{s}}^{b_1}$ and $E_{\mathrm{s}}^{b_2}$, are well known [349]. They are given by the right-hand side of (3.41) if the cutoff radius r_{c_i} is replaced by b_1 and b_2, respectively.

The energies of interaction between grain boundary dislocations of the pile-up and climbing grain boundary dislocations were calculated in the following way [428]. First the energy of interaction between the climbing dislocation and an arbitrary dislocation of the pile-up was found. This procedure is quite cumbersome because it involves determination of the shear stress created by the climbing dislocation, which acts in the plane of the pile-up, with further calculation of the work done to generate a dislocation from the pile-up in this shear stress field. For example, the energy of interaction of the $\boldsymbol{b}_1$-dislocation with the i th dislocation from the pile-up was obtained as [428]

$$E_i^{b-b_1} = Dbb_1 \sin\phi \left(\frac{p^2 + Rp\cos\phi}{R^2 + 2Rp\cos\phi + p^2} - \frac{p^2 + x_i p \cos\phi}{x_i^2 + 2x_i p\cos\phi + p^2} \right.$$
$$\left. + \frac{1}{2} \ln \frac{R^2 + 2Rp\cos\phi + p^2}{x_i^2 + 2x_i p \cos\phi + p^2} \right) , \qquad (4.47)$$

where 2ϕ is the angle between the grain boundaries along which the $\boldsymbol{b}_1$- and $\boldsymbol{b}_2$-dislocations climb, p is the path of the $\boldsymbol{b}_1$-dislocation, and x_i is the coordinate of the i th dislocation in the pile-up (Fig. 4.28b). The energy $E_i^{b-b_2}$ of interaction of the $\boldsymbol{b}_2$-dislocation with the i th dislocation in the pile-up is calculated in a similar way. Then the total energy characterizing the interaction of both the $\boldsymbol{b}_1$- and $\boldsymbol{b}_2$-dislocations with grain boundary dislocations of the pile-up reads [428]

$$E_{\Sigma 2}^{b-b_1} + E_{\Sigma 2}^{b-b_2} = \sum_{i=1}^{n_{\mathrm{c}}} \left(E_i^{b-b_1} + E_i^{b-b_2} \right) . \qquad (4.48)$$

The energy $E^{b_1-b_2}$ of interaction between the climbing dislocations is also calculated as the work done to generate one dislocation in the stress field of the other. Omitting intermediate results, we give the final formula [428]:

$$E^{b_1-b_2} = -\frac{Db_1b_2}{2} \left[\frac{2R^2}{R^2 \mathrm{cosec}^2\phi - 4Rp\,\mathrm{ctg}\phi + 4p^2} \right.$$
$$\left. + \cos(2\phi) \ln \frac{R^2 - 2Rp\sin(2\phi) + 4p^2\sin^2\phi}{4p^2\sin^2\phi} \right] . \qquad (4.49)$$

The work $E^{\tau - b_k}$ done to transfer the $\boldsymbol{b}_k$-dislocation through a distance p under the shear stress τ is equal to

$$E^{\tau - b_k} = -b_k p \tau \sin(2\phi) \ . \tag{4.50}$$

We now have all terms featuring in the energy difference $\Delta W = W_2 - W_1$ that characterizes the splitting (Fig. 4.28) treated as an elementary act of the crossover from grain boundary sliding to rotational deformation mode in NCMs. The numerical results for the dependences $\Delta W(p)$, obtained for $b_1 = b_2 = b/(2\sin\phi)$, $n_c = 5$, $R = 10^7 b$, $\tau = G/200$ and different values of the angle 2ϕ, are presented in Fig. 4.29 [428]. These curves show that the splitting – an elementary act of the crossover from grain boundary sliding to rotational mode – is energetically favorable at large values (80°–120°) of the angle 2ϕ. In [309], it has been demonstrated that the splitting of the head grain boundary dislocation of a pile-up stopped at a triple junction into two, gliding and sessile, grain boundary dislocations is energetically favorable at low values of the angle 2ϕ. This version of the splitting serves as an elementary act of grain boundary sliding at triple junctions.

Thus the crossover from grain boundary sliding to rotational deformation (Figs. 4.27 and 4.28) occurs effectively at triple junctions with large values of the angle 2ϕ, while the grain boundary sliding itself occurs effectively at triple junctions with comparatively low values of the angle 2ϕ.

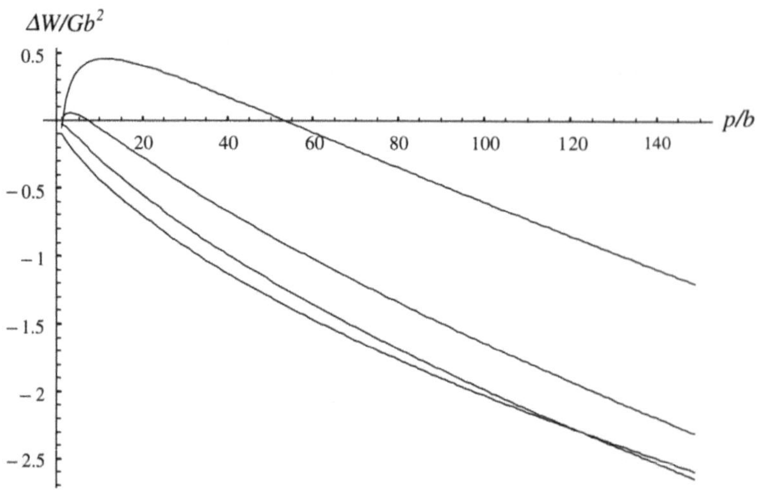

Fig. 4.29. Dependence of ΔW (in units of Gb^2) on distance p (in units of b) moved by climbing $\boldsymbol{b}_1$- and $\boldsymbol{b}_2$-dislocations, for $2\phi = 40, 60, 80$ and 120° (*from top to bottom*)

4.2.3.2 Cooperative Action of Grain Boundary Sliding and Rotational Deformation Mode in Nanocrystalline Materials. (Large-Scale View)

Consider how the system of defects described in the previous section develops in time under an external shear stress [428]. After the head grain boundary dislocation has split into the $\boldsymbol{b}_1$- and $\boldsymbol{b}_2$-dislocations (Fig. 4.28b), their stress fields prevent movement of the second grain boundary dislocation of the pile-up towards the triple junction where the splitting has occurred. When the climbing $\boldsymbol{b}_1$- and $\boldsymbol{b}_2$-dislocations move far from the triple junction, their effect on the second grain boundary dislocation of the pile-up becomes weak. In this case the second dislocation of the pile-up moves to the triple junction where it splits into two grain boundary dislocations climbing along the adjacent grain boundaries. Such a splitting process occurs repeatedly, transforming grain boundary dislocations of the pile-up into climbing grain boundary dislocations (Fig. 4.26). These climbing dislocations form dislocation walls of finite extent at the two grain boundaries adjacent to the triple junction. The geometric and elastic properties of these two ragged dislocation walls are similar to those of the corresponding two-axis dipoles of partial wedge disclinations [56, 263, 397, 398].

Now assume that similar splitting processes occur at two opposite triple junctions of grain boundaries surrounding a nano-sized grain. The grain boundary dislocation pile-ups located near the opposite triple junctions consist of dislocations with Burgers vectors of opposite signs, which tend to move towards each other under the shear stress action and are stopped by the triple junctions. Such a pile-up consisting of n_c dislocations in its initial state may be represented effectively as a superdislocation with Burgers vector $\pm\boldsymbol{B} = \pm n_c\boldsymbol{b}$ (Fig. 4.30a). Following the model [428], these two superdislocations split and are transformed into four grain boundary dislocation walls of finite extent, which are effectively represented as four dipoles of disclinations (Fig. 4.30b).

In calculating the energy characteristics of evolution of the grain boundary dislocation ensemble, causing grain rotation, for definiteness and simplicity, the authors [428] have made the following model assumptions:

- The grain is a hexagon and the angles 2ϕ characterizing the splitting processes at the opposite triple junctions are the same.
- The magnitudes of the Burgers vectors of all grain boundary dislocations belonging to dislocation pile-ups are the same.
- The plane of the shear stress action is parallel to planes of grain boundaries where the grain boundary dislocation pile-ups are located.
- All the disclinations modeling the walls of the climbing grain boundary dislocations have the same strength magnitude ω. Furthermore, the total number $2n_c$ of grain boundary dislocations participating in the transformations is believed to be constant.

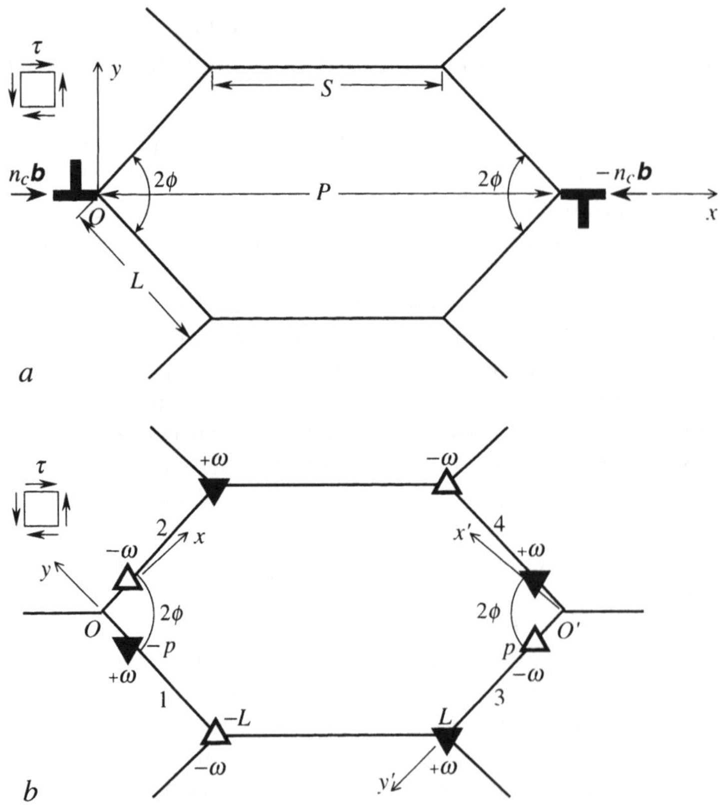

Fig. 4.30. Large-scale model of crossover from grain boundary sliding to rotational deformation mode. (**a**) Two superdislocations (models of pile-ups of grain boundary dislocations) are stopped at the opposite triple junctions. (**b**) Four disclination dipoles (models of climbing dislocation walls) are located at the grain boundaries adjacent to the triple junctions

It is worth noting that none of these assumptions are necessary and any of them may be omitted in a more general model.

Following [428], let us consider the conditions required to realize the total transformation of two initial pile-ups of grain boundary dislocations (Fig. 4.30a) into four ragged walls of climbing grain boundary dislocations modeled by four dipoles of grain boundary disclinations (Fig. 4.30b). The transformation of the grain boundary dislocation ensemble may be characterized by the difference $\Delta\widetilde{W} = \widetilde{W}_2 - \widetilde{W}_1$ between the energies of the final ($\widetilde{W}_2$) and initial ($\widetilde{W}_1$) states of the ensemble. The transformation is energetically favorable (unfavorable), if $\Delta\widetilde{W} < 0$ ($\Delta\widetilde{W} > 0$).

In the framework of the model discussed, the energy of the dislocation ensemble in its initial state (Fig. 4.30a) consists of the three terms [428]

$$\widetilde{W}_1 = 2E_1^{\text{pu}} + 2E_{\Sigma 1}^{b-b} + E_{\text{int}}^{B-B} , \tag{4.51}$$

where the two first terms are already known – they describe the sum of the self-energies of the grain boundary dislocations in the pile-ups ($2E_1^{\text{pu}}$) and the sum of all energies of pairwise interactions between the grain boundary dislocations within their pile-ups ($2E_{\Sigma 1}^{b-b}$). They are determined by (4.40) and (4.45), respectively. The third term, E_{int}^{B-B}, is the energy that characterizes pairwise interactions between grain boundary dislocations belonging to the different pile-ups. In the first approximation, this energy may be estimated as the energy of elastic interaction of two superdislocations with Burgers vectors $\pm\boldsymbol{B}$ [428]:

$$E_{\text{int}}^{B-B} = -DB^2 \ln \frac{R}{P} , \tag{4.52}$$

where P is the distance between the opposite triple junctions

The energy (4.51) of the system in its initial state is thus approximated by the sum of three terms. The first two terms strictly take into account the fine structure of the dislocation pile-ups (i.e., the specific distributions of dislocations within the pile-ups), while the third is approximated by (4.52), which does not account for this fine structure. However, this approximation (when a pile-up is treated as a superdislocation) gives rather good results if the distance P (Fig. 4.30a) is much larger than the pile-up length [349]. The latter must be smaller than $S/2$, and S is assumed always to be smaller (two or more times, see below) than P in the model [428]. Therefore, the pile-up length must be four or more times smaller then P, thereby justifying the superdislocation approximation in calculating the term (4.52).

The energy of the grain boundary dislocation ensemble in its final state (Fig. 4.30b) comprises the six terms [428]

$$\widetilde{W}_2 = 4E_{\text{s}}^{\triangle} + 4E_{\tau}^{\triangle} + 2E_{\tau}^{\text{pu}} + 2E_{\text{int}}^{1-2} + 2E_{\text{int}}^{1-3} + 2E_{\text{int}}^{1-4} , \tag{4.53}$$

where $E_{\text{s}}^{\triangle}$ denotes the self-energy of a disclination dipole, $E_{\tau}^{\triangle}$ the work done to transfer the grain boundary dislocations resulting from the splitting under the shear stress τ, E_{τ}^{pu} the work done to transfer all grain boundary dislocations belonging to the pile-ups from their initial positions up to the corresponding triple junctions under the shear stress τ, and E_{int}^{i-j} the energy of interaction between the ith and jth disclination dipoles ($i \neq j,\ i = 1, 2, 3,\ j = 2, 3, 4$).

The total self-energy $E_{\text{s}}^{\triangle}$ of the disclination dipole with arm $(L - p)$ and strength ω is given by (4.17), replacing L by $(L - p)$, N by n_{c}, and b by b_1.

The shear stress τ acts on the climbing grain boundary dislocations which compose the disclination dipoles. The work $E_{\tau}^{\triangle}$ done to climb of n_{c} dislocations within one disclination dipole is [428]

$$E_{\tau}^{\triangle} = -\tau b_1 \sin 2\phi \left[p + \frac{n_{\text{c}}(n_{\text{c}} - 1)b_1}{2\omega} \right] . \tag{4.54}$$

This work is the same for all the disclination dipoles under consideration (Fig. 4.30b).

The grain boundary dislocations of each pile-up subsequently move under the action of the shear stress τ from their initial positions $x = x_i$ to the new positions at the triple junctions ($x = 0$). The work E_τ^{pu} done to transfer all the grain boundary dislocations in a pile-up is [428]

$$E_\tau^{\mathrm{pu}} = -\tau b \sum_{i=1}^{n_c} |x_i| \,. \tag{4.55}$$

The energy E_{int}^{i-j} of elastic interaction between the ith and jth disclination dipoles is calculated as the work done to generate one dipole in the stress field of the other. Omitting cumbersome intermediate calculations, we give the final results [428]:

$$E_{\mathrm{int}}^{1-2} = \frac{D\omega^2}{4}\Big[\Psi_-(x',L,0,0,y',\phi)-\Psi_-(x',p,0,0,y',\phi)\Big]\Big|_{x'=0}^{x'=R}\Big|_{y'=p}^{y'=L}, \tag{4.56}$$

$$E_{\mathrm{int}}^{1-3} = \frac{D\omega^2}{4}\Big[\Psi_+(x',L,T,T',y',\phi)-\Psi_+(x',p,T,T',y',\phi)\Big]\Big|_{x'=0}^{x'=R}\Big|_{y'=p}^{y'=L}, \tag{4.57}$$

$$E_{\mathrm{int}}^{1-4} = \frac{D\omega^2}{4}\Big[\Psi_+(x',T',T,L,y',0)-\Psi_+(x',T',T,p,y',0)\Big]\Big|_{x'=0}^{x'=R}\Big|_{y'=p}^{y'=L}, \tag{4.58}$$

where

$$\begin{aligned}\Psi_\pm(x',l,t,t',y',\phi) = {} & \Big[(x'+t)^2 + l^2 + (y'-t')^2 \\ & \pm 2l(y'-t')\cos(2\phi) \mp 2l(x'+t)\sin(2\phi)\Big] \\ & \times \ln\Big[(x'+t)^2 + l^2 + (y'-t')^2 \\ & \pm 2l(y'-t')\cos(2\phi) \mp 2l(x'+t)\sin(2\phi)\Big]\end{aligned} \tag{4.59}$$

and $T = P\sin\phi$, $T' = P\cos(2\phi)$.

We have thus identified all terms featuring in (4.51) for the energy $\widetilde{W}_1$ and (4.53) for the energy $\widetilde{W}_2$, which compose the characteristic difference $\Delta\widetilde{W} = \widetilde{W}_2 - \widetilde{W}_1$. If $\Delta\widetilde{W}$ is negative (positive), the transformation of the defect configuration is energetically favorable (unfavorable). The authors [428] have analysed numerically the dependence of $\Delta\widetilde{W}$ on the disclination dipole strength ω (Fig. 4.31). The calculations have been carried out for $b_1 = b_2 = b/(2\sin\phi)$, $n_c = 5$, $R = 10^7 b$, $\tau = G/200$, $L = 200b$, $P = 400b$ and different values of the angle 2ϕ that characterizes the geometry of the triple junction. The plots $\Delta W(\omega)$ show that the transformation is most favorable in the range of 2ϕ from $100°$ to $160°$. In addition, the magnitude of $\Delta\widetilde{W}$ (that is, the energy gain due to the transformation when $\Delta\widetilde{W} < 0$) decreases with an increase in the disclination strength ω (ranging from 0.05 to 0.1).

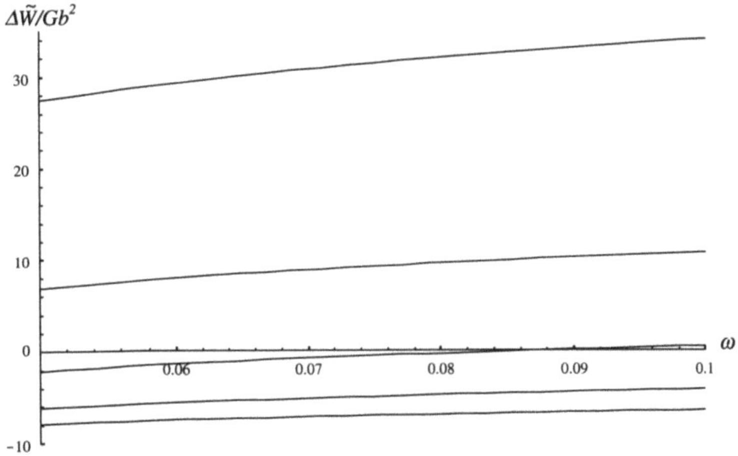

Fig. 4.31. Dependence of $\Delta\widetilde{W}$ (in units of Gb^2) on the disclination strength ω in the range of values of the triple junction angle $2\phi = 60$, 80, 100, 120 and 160° (*from top to bottom*)

4.2.3.3 Critical Stress of Crossover from Grain Boundary Sliding to Rotational Deformation

The equation $\Delta\widetilde{W} = 0$ allows one to calculate the critical shear stress τ_c at which the crossover from grain boundary sliding to rotational deformation occurs (Fig. 4.30). This equation may be written in the form [428]

$$\tau_c = \frac{\widetilde{W}_2' - \widetilde{W}_1(\tau_c)}{2b\sum_{i=1}^{n_c} |x_i(\tau_c)| + 4b_1 \sin 2\phi \left[p + \dfrac{n_c(n_c-1)b_1}{2\omega}\right]}\,, \tag{4.60}$$

where $\widetilde{W}_2'$ is given by (4.53) for $\widetilde{W}_2$ with terms $4E_\tau^\Delta$ and $2E_\tau^{\mathrm{pu}}$ removed.

Let us consider how τ_c changes depending on the values of the angle 2ϕ and grain size P. The authors [428] used the following characteristic values of parameters of an NCM and the defect configuration under consideration. The Poisson ratio ν was 0.3. The moduli of the grain boundary dislocation Burgers vectors and disclination strength were taken as $b = 0.1$ nm and $\omega = 0.1$ ($\approx 5.7°$). The number n_c was varied from 3 to 20, together with the distance L, to keep the disclination strength ω constant. The distance p was also assumed to be constant and equal to l. These assumptions are considered reasonable on the basis of the information available. As we will see later, the actual form of the curves is not very sensitive to the values.

We now express the distance L in terms of S and P, whose ratio $q = S/P < 1$. We then obtain

$$L = \frac{1-q}{2\cos\phi} P\,. \tag{4.61}$$

With the formulas under discussion and for the above characteristic values of the parameters, the dependence of the critical shear stress τ_c on the grain size (diameter) P was calculated in [428]. This dependence is shown for $q = 0.5$ (Fig. 4.32a) and $q = 0.1$ (Fig. 4.32b), for different values of the triple junction angle 2ϕ. The curves $\tau_c(P)$ lie, in general, in different ranges of P because the

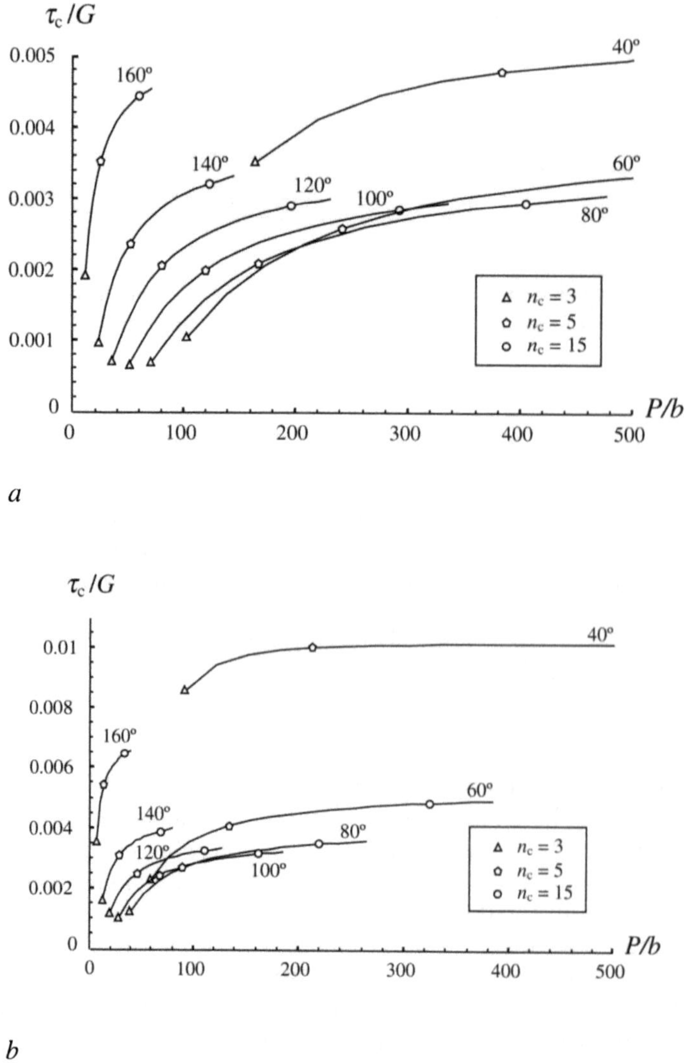

Fig. 4.32. Dependence of the critical shear stress τ_c on the grain size P, (**a**) for $q = 0.5$ and (**b**) for $q = 0.1$, for different values of the triple junction angle 2ϕ given by the *figures on the curves*. The numbers of grain boundary dislocations n_c are shown by *triangles* ($n_c = 3$), *pentagons* ($n_c = 5$) and *circles* ($n_c = 15$)

disclination strength ω is taken to be constant here. Therefore an increase in P leads to an increase in both L and n_c, and this relation depends on the angle 2ϕ. For the sake of convenience, the points corresponding to $n_c = 3$ (triangles), 5 (pentagons) and 15 (circles) are attached to every curve. The upper ends of the curves correspond to $n_c = 20$. As a result, one can also trace how τ_c depends on the angle 2ϕ when both L and n_c are kept constant.

From Fig. 4.32, it follows that τ_c decreases with decreasing P (grain refinement) [428]. This is the main result of this section. It demonstrates that smaller grains are rotated more easily (i.e., they are rotated under the action of smaller critical stress τ_c) than the larger ones. Moreover, τ_c strongly depends on the grain shape (i.e., on q and 2ϕ). An increase in q leads to a decrease in τ_c, because larger q (for fixed values of P and 2ϕ) requires a smaller number n_c of grain boundary dislocations to spread along the grain boundary (or smaller dipole arm $L - p$). The dependence of τ_c on the angle 2ϕ is more complicated. For fixed L and n_c, in the range of relatively small angles, $2\phi < 100°$, the critical stress τ_c decreases with increasing 2ϕ and achieves its minimum at $2\phi \approx 100°$. In the range of relatively large angles, $2\phi > 100°$, the critical stress τ_c increases with rising 2ϕ.

One may thus conclude that smaller grains characterized by larger q and $2\phi \approx 100°$ require smaller critical stress τ_c to rotate [428].

The theoretical models for motion of grain boundary disclinations considered in Sect. 4.2 allow us to draw the following conclusions:

- Motion of grain boundary disclinations is an effective mechanism of rotational deformation in NCMs.
- Motion of grain boundary disclinations may be realized through the micromechanisms of emission of lattice dislocations or climb of grain boundary dislocations. Realization of these micromechanisms is possible for a wide range of values of the main geometry and force parameters of the system.
- Motion of grain boundary disclinations by means of lattice dislocation emission is effective until the usual dislocation plasticity is possible in nanograins. When the glide of lattice dislocations is replaced by grain boundary sliding, the appearance of grain boundary dislocation pile-ups at grain boundary triple junctions and further transformation of these pile-ups into the walls of climbing grain boundary dislocations (the dipoles of grain boundary disclinations) serve as a possible way to activate the rotational mode of plastic deformation in NCMs.
- The transition from grain boundary sliding to rotational deformation becomes energetically favorable when the external shear stress achieves its critical value, which depends on the elastic properties of the NCM, the structure of its grain boundaries, its grain size and its shape. Smaller grains require smaller critical stress to rotate.

5 Disclinations, Amorphization and Cracks at Grain Boundaries in Nanocrystalline Materials

The processes developing in the late stages of plastic deformation in conventional (i.e., non-nanocrystalline in their initial state) crystalline materials are characterized by defect structures which have passed over some intermediate steps in their evolution [56, 263–266, 381, 389, 430–440]. Their principal difference from the defect structures inherent in the initial stages of plastic deformation is the mainly disclinational (rotational) character of the defect configurations (misorientation bands, fragment boundaries, high-angle grain boundaries of deformation origin, etc.) observed in experiments, which may be considered as carriers of rotational plasticity. At this late stage of plastic deformation in metallic materials, there exist two general paths for further evolution of the defect structures. The first path is the generation of microcracks in the regions of highest stress concentration, leading to fracture of the material. The second alternative is a transition to a nanocrystalline and/or amorphous structure of the metal [441, 442]. Obviously, the second path for structural evolution allows the deformed sample to achieve larger steps of plastic deformation than the first and this explains why it is very promising for metal-forming technologies. Nowadays, this approach is widely used in various techniques for fabricating NCMs and amorphous alloys (e.g., ball milling and mechanical alloying of powders [14], equal-channel angular pressing [28,443], etc.). In the 1980s, probably the first amorphous–nanocrystalline composites [36,37] were obtained under the combined action of high pressure and intensive shear.

In order to address the question as to which of the two evolutionary channels is expected to be realized in mechanically loaded crystalline materials, one should first study micromechanisms of amorphization and fracture processes in such materials. In this section, we discuss theoretical studies [206, 256, 425–427] investigating the splitting of triple junction and grain boundary disclinations into ensembles of disclinations of lower strength, which can lead to local amorphization of the material structure and reduce the probability of microcrack generation. We also consider a recent model [444], in which nanocracks are assumed to nucleate and grow along grain boundaries, changing their direction of growth at triple junctions of grain boundaries. We focus on the triple junction and grain boundary disclinations for three reasons:

- Such defects are typical representatives of the defect structures in the late stages of plastic deformation in conventional polycrystals [56,263–266,431] and in the initial structure of NCMs (see Sect. 4.2).
- They are powerful sources of elastic stresses [56,263,397,398] and favorable sites for microcrack generation [56, 263, 264, 266, 430, 431].
- They are natural structural elements of the amorphous state [8, 56, 260–263, 267, 270, 445, 446] and can play an embryonic role for the amorphous phase.

On the basis of disclination models for the structure of amorphous, nanocrystalline and severely deformed polycrystalline solids, one can therefore study both the aforementioned evolutionary channels (microcrack generation and local solid-state amorphization) under large deformations.

5.1 Splitting of a Triple Junction Disclination and Amorphization of the Triple Junction

Disclinations at triple junctions of grain boundaries (triple junction disclinations) have been studied during almost three decades [56, 180, 263, 264, 304, 447–453]. Here we consider one of the important features of triple junction disclinations, namely their capacity to split into disclinations of lower strength [56, 263]. It follows from elastic models for disclinations that their splitting is energetically favorable because their strain energy is proportional to the square of their strength, $W \sim \Omega^2$, and therefore the total strain energy of the split disclination configuration is smaller than that of the initial dislocation (as is the case when perfect dislocations split into partials [349]). For example, Gryaznov et al. [88] have shown that the splitting of a wedge disclination into two disclinations with the same total strength can serve as one of the main channels for elastic stress relaxation in pentagonal nanoparticles.

The scheme whereby a triple junction disclination splits into several disclinations of smaller strength with the appearance of an embryo of the amorphous phase at the triple junction (Fig. 5.1) has been described qualitatively in [205, 454]. The driving force for the splitting process is a decrease in the disclination strain energy which accompanies the splitting. At the same time, there is a hampering force for the splitting. This is related to the formation of the local amorphous phase whose specific energy is usually higher than that of the initial crystalline phase [8, 56, 260–263, 267, 445, 446]. The splitting occurs when the driving force reaches (or becomes larger than) the value of the hampering force. The main aim of this section is to define quantitative conditions under which disclination splittings associated with local amorphization at triple junctions of grain boundaries occur in polycrystalline solids.

Consider a straight wedge disclination with strength Ω, placed at a triple junction of grain boundaries, which splits into an ensemble of small-strength

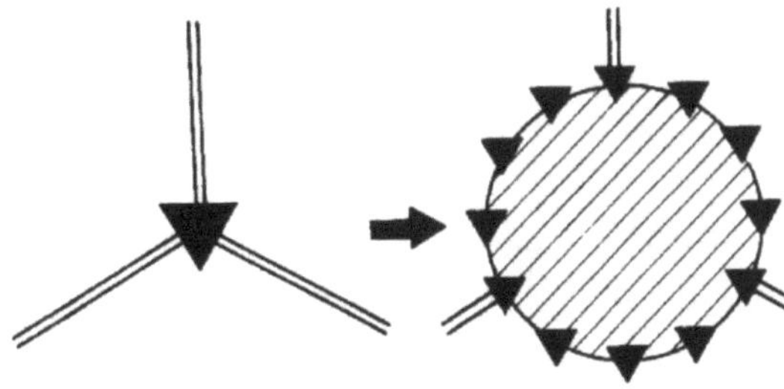

Fig. 5.1. Splitting of a triple junction disclination (**a**) into several small-strength disclinations (**b**), in which case an amorphous embryo (*shaded region*) is formed

disclinations with the same total strength, periodically distributed along a circle with center at the triple junction (Fig. 5.1). Both the strain self-energy of a disclination and the energy of the disclination–disclination interaction are known to depend strongly on boundary conditions [56, 263]. Hereafter, for definiteness, we confine ourselves to analysis of the disclination ensemble located in an infinitely long cylinder with free (unloaded) surface [206].

5.1.1 Energy of Wedge Disclinations in a Cylinder

Calculations of the total energy of the ensemble of wedge disclinations in a cylinder have been carried out using the energies of their pair interaction. The pair interaction energy for wedge disclinations in a cylinder with a free surface was obtained by Romanov [54] (see also [56, 263]). For two disclinations with strengths ω_i and ω_j, whose lines are shifted from the axis of the cylinder of radius R by the distances R_i and R_j (Fig. 5.2), respectively, this energy (per unit length of disclinations) may be written in the compact form [206]

$$W^{(i,j)} = \frac{D\omega_i\omega_j}{4}\left(l_{ij}^2 + d_{ij}^2 \ln\frac{d_{ij}^2}{l_{ij}^2 + d_{ij}^2}\right), \tag{5.1}$$

where $D = G/[2\pi(1-\nu)]$, d_{ij} is the distance between the disclinations ($d_{ij}^2 = R_i^2 + R_j^2 - 2R_iR_j\cos\theta_{ij}$), and $l_{ij}^2 = R^2 - R_i^2 - R_j^2 + R_i^2R_j^2/R^2$.

Accordingly, the total elastic energy of N parallel disclinations in the cylinder is defined by

$$W^{(N)} = \frac{1}{2}\sum_{i=1}^{N}\sum_{j=1}^{N} W^{(i,j)} = \frac{D}{8}\sum_{i=1}^{N}\sum_{j=1}^{N}\omega_i\omega_j\left(l_{ij}^2 + d_{ij}^2\ln\frac{d_{ij}^2}{l_{ij}^2 + d_{ij}^2}\right). \tag{5.2}$$

The sum (5.2) contains terms with $i \neq j$, each being the energy of the interaction between two (different) disclinations, and terms with $i = j$, each being twice the self-energy of the ith disclination.

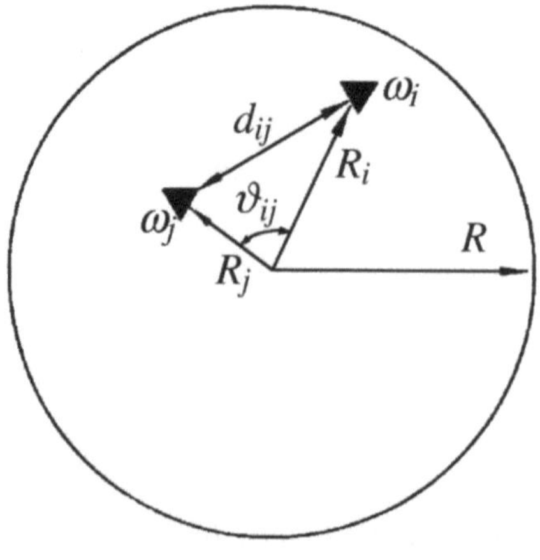

Fig. 5.2. Wedge disclinations in a free-surface cylinder

5.1.2 Necessary Conditions for Disclination Splitting

Let us use (5.2) to estimate the gain in the strain energy due to the splitting if an initial triple junction disclination of strength $\Omega = N\omega$ lying along the cylinder axis splits into an ensemble of N disclinations of strength ω lying at a distance R_0 from the cylinder axis (Fig. 5.1). In such a situation, one obtains parameters: $d_{ij}^2 = 2R_0^2(1-\gamma_{ij})$, $l_{ij}^2 = R^2(1-R_0^2/R^2)^2$, and $\gamma_{ij} = \cos\theta_{ij}$. Substituting these parameters into (5.2), we find

$$W^{(N)} = \frac{D(N\omega)^2}{8}R^2 - \frac{D\omega^2 R_0^2}{4}\left\{N^2\left(1-\frac{R_0^2}{2R^2}\right)\right.$$
$$\left. + \sum_{i=1}^{N}\sum_{j=1}^{N}(1-\gamma_{ij})\ln\left[1+\frac{(R^2-R_0^2)^2}{2(1-\gamma_{ij})R^2R_0^2}\right]\right\}, \tag{5.3}$$

where the first term on the right-hand side of (5.3) represents the strain self-energy of the initial disclination with strength $N\omega$, whereas the second term is the gain ΔW in the total energy of the system, related to the splitting of the initial disclination into N small-strength (ω) disclinations. For $R_0 \ll R$, the gain is

$$\Delta W|_{R_0\ll R} \approx -\frac{D\omega^2R_0^2}{4}\left[N^2+\sum_{i=1}^{N}\sum_{j=1}^{N}(1-\gamma_{ij})\ln\frac{R^2}{2(1-\gamma_{ij})R_0^2}\right]. \tag{5.4}$$

Let us examine the dependence of $\Delta W|_{R_0\ll R}$ on the number N of resulting disclinations when the disclination strength ω is fixed. In doing so, we represent $\Delta W|_{R_0\ll R}$ in the form

$$\Delta W(N)|_{R_0 \ll R} \approx -\frac{D\Omega^2 R_0^2}{4}\Psi(N)\,, \tag{5.5}$$

where the function $\Psi(N)$ is defined by

$$\Psi(N) = 1 + \frac{1}{N^2}\sum_{i=1}^{N}\sum_{j=1}^{N}(1-\gamma_{ij})\ln\frac{R^2}{2(1-\gamma_{ij})R_0^2}\,. \tag{5.6}$$

If the resulting disclinations are periodically distributed along the circle then $\gamma_{ij} = \cos[2\pi(i-j)/N]$, and direct calculation of the sum (5.6) shows that, for $N > 3$, the dependence $\Psi(N)$ is very weak and can be approximated by the constant $\Psi \approx 2\ln(R/R_0)$ [206]. Then

$$\Delta W(R_0 \ll R,\ N \gg 1) \approx -\frac{D}{2}\Omega^2 R_0^2 \ln\frac{R}{R_0}\,. \tag{5.7}$$

In such a situation when the resulting disclinations are periodically distributed along the circle with its center at the cylinder axis, the gain in strain energy is practically independent of the disclination number but strongly depends on both the splitting radius R_0 and the cylinder radius R (size of the sample). As these radii grow, so does the energy gain ΔW. In other words, the gain ΔW in the strain energy of the disclination ensemble serves as an effective driving force for the splitting of a triple junction disclination (Fig. 5.1).

We now analyse the hampering force for the splitting. Such splitting is accompanied by the formation of the local amorphous phase – the spread core of the split disclination (see Fig. 5.1b). The energy density of the new amorphous phase usually exceeds that of the initial crystalline phase due to the disorder of the atomic structure in the amorphous phase [8, 56, 260–263, 267, 445, 446]. This causes the basic hampering force for the splitting of disclinations. As to details, the energy of the amorphous embryo (per unit length of disclinations) may be estimated as $W_{\mathrm{am}} = \pi R_0^2 \varepsilon_{\mathrm{a-c}}$, where $\varepsilon_{\mathrm{a-c}}$ is the difference between the energy densities of the amorphous and crystalline phases.

A necessary condition for the splitting and local amorphization is therefore the inequality $|\Delta W| \geq W_{\mathrm{am}}$. As a result, the balance of the driving and hampering contributions, $\Delta W + W_{\mathrm{am}} = 0$, gives the critical strength of the initial disclination [206]:

$$\Omega_{\mathrm{c}} = \sqrt{\frac{2\pi\varepsilon_{\mathrm{a-c}}}{D\ln(R/R_0)}}\,. \tag{5.8}$$

If the triple junction disclination has strength $\Omega > \Omega_{\mathrm{c}}$, then it can split into a number of small-strength disclinations, accompanied by the formation of the amorphous embryo at the triple junction. One can estimate the critical strength Ω_c for some typical situations which differ in the characteristic value of ε_{a-c}.

Let the initial crystalline phase contain a small or intermediate density of defects. Such a situation is rather normal. In these circumstances, the difference ε_{a-c} between the energy densities of the amorphous and crystalline phases is high and ranges from $G/83$ to $G/65$ [270]. Then, for typical values of $\varepsilon_{a-c} \approx G/83$ to $G/65$, $\nu = 0.3$, (5.8) implies $\Omega_c \approx \pi/8$ to $\pi/7$, $\pi/11$ to $\pi/10$, $\pi/14$ to $\pi/12$, $\pi/17$ to $\pi/15$, and $\pi/18$ to $\pi/16$, if the ratio $R/R_0 = 10, 10^2, 10^3, 10^4$, and 10^5, respectively. These values of Ω_c are rather high. Triple junction disclinations in real polycrystalline metallic materials usually show strengths $\leq \pi/60$ [264, 448, 452], smaller than Ω_c. Such disclinations cannot split, inducing local amorphization in the situation discussed here.

High-density ensembles of defects are usually present in polycrystalline solids under high-strain deformation. For example, dislocation ensembles with limiting densities $\geq 10^{16}$ m^{-2} are characteristic elements of the defect structures of metallic solids at high strains, being close to strains at which deformation-induced amorphization occurs [455]. The presence of high-density ensembles of defects essentially increases the energy density of the crystalline phase, in which case this density becomes close to the energy density of the amorphous phase. As a result, ε_{a-c} in solids with high densities of defects is essentially smaller than ε_{a-c} in solids with small or intermediate densities of defects. According to (5.8), this decreases the critical strength Ω_c for splitting of triple junction disclinations. For instance, for $\varepsilon_{a-c} = 10^{-3}G$, $\nu = 0.3$, and $R/R_0 = 10^5$, (5.8) yields $\Omega_c \approx \pi/60$. Triple junction disclinations with this strength have been observed in conventional polycrystalline metals [264,448,453,456]. As (5.8) implies, these real disclinations are able to split into small-strength disclinations (with local amorphous phase formed at the corresponding triple junctions) in metallic materials with high densities of defects – say, materials under high-strain deformation.

One can conclude that the splitting of experimentally observed triple junction disclinations, accompanied by local solid-state amorphization, is only possible under severe plastic deformation, when the density of defects in the vicinity of the triple junctions is very high.

It is worth noting that realization of the considered scheme of disclination splitting can be greatly facilitated in some special cases, when the difference between the energy densities of the amorphous and crystalline states ε_{a-c} becomes negative [205, 206, 454].

5.1.3 Microcrack Generation at an Amorphized Triple Junction

In the last section we showed that amorphization at a grain boundary triple junction, related to the splitting of a triple junction disclination (Fig. 5.1), leads to a decrease in the latent elastic energy in a solid. This allows one to think of the above amorphization as an alternative relaxation process (under

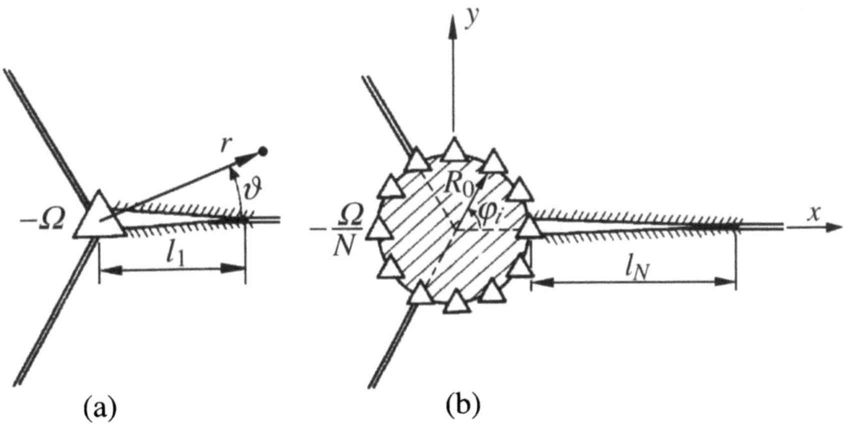

Fig. 5.3. Microcrack generation at triple junctions of grain boundaries. (**a**) Generation of a microcrack with equilibrium length l_1 at a triple junction with initial (unsplit) negative wedge disclination of strength $-\Omega$. (**b**) Generation of a microcrack with equilibrium length l_N at one of the $-\Omega/N$-strength disclinations bordering the amorphized region of radius R_0 at the triple junction. Both microcracks (**a**) and (**b**) are generated along one of the grain boundaries

special conditions) to 'normal' relaxation processes such as plastic deformation and generation of microcracks. In these circumstances, amorphization decreases the probability of microcrack generation at grain boundary junctions. This effect is characterized quantitatively as the amorphization-induced increase of the 'equilibrium' length of a microcrack. (In general, as the 'equilibrium' length grows, the probability of microcrack generation decreases.)

Rybin and Zhukovskii [448] suggested a theoretical model which describes microcrack generation at a loop of triple junction disclinations. In [206], we produced similar considerations for the case of a straight wedge disclination (Fig. 5.3a) and extended this model to the situation with a split disclination configuration bordering a cylindrical embryo of amorphous phase (Fig. 5.3b). Following [206], we will calculate and compare the equilibrium lengths which characterize microcracks generated at disclinated (Fig. 5.3a) and amorphized (Fig. 5.3b) triple junctions of grain boundaries.

First, let us consider a straight wedge disclination with negative strength $-\Omega$, located along the central axis of an isotropic elastic cylinder with radius R (Fig. 5.3a). The tangential component of the stress field created by the disclination is [56, 263]

$$\sigma_{\theta\theta}^{(1)} = D\Omega \left(\ln \frac{R}{r} - 1 \right) . \qquad (5.9)$$

The tensile stresses are capable of inducing the generation of a microcrack along one of the three grain boundaries. To calculate the equilibrium length

l_1 of such a microcrack, we use the well-known configurational force method [457].

Hence, following Indenbom [457], the configurational force F, which is defined as the strain energy released when the microcrack moves over a unit distance, can be written as

$$F = \frac{\pi(1-\nu)l_1}{4G}[\bar{\sigma}_{\theta\theta}^{(1)}]^2 , \tag{5.10}$$

where $\bar{\sigma}_{\theta\theta}^{(1)}$ is the mean weighted stress acting on the segment $r \in (0, l_1)$, $\theta = 0$ before the microcrack has been generated:

$$\bar{\sigma}_{\theta\theta}^{(1)} = \frac{2}{\pi l_1}\int_0^{l_1} \sigma_{\theta\theta}^{(1)}(r)\sqrt{\frac{r}{l_1 - r}}\mathrm{d}r . \tag{5.11}$$

The standard condition $F = 2\gamma$, where γ is the specific surface energy, yields the following energetic condition for equilibrium of the microcrack front:

$$\bar{\sigma}_{\theta\theta}^{(1)} = \sqrt{\frac{8G\gamma}{\pi(1-\nu)l_1}} . \tag{5.12}$$

Substituting (5.9) into (5.11), after some algebra, one finds

$$\bar{\sigma}_{\theta\theta}^{(1)} = D\Omega\left(\ln\frac{R}{l_1} + 2\ln 2 - 2\right) . \tag{5.13}$$

Then (5.12) and (5.13) imply [206]

$$\frac{32\pi\gamma(1-\nu)}{G\Omega^2} = l_1\left(\ln\frac{R}{l_1} + 2\ln 2 - 2\right)^2 , \tag{5.14}$$

which allows one to calculate the critical crack length l_1. Hence, for typical values of $\Omega \approx 0.03$ [448], $\gamma \approx Gb/8$, $b \approx 0.3$ nm, $\nu \approx 0.3$, $R \approx 10^{-4}$ m, (5.14) yields $l_1 \approx 67$ nm. This value of l_1 is close to the value $l_1 \approx 70$ nm estimated by Rybin and Zhukovskii [448], and to the experimentally observed length $l_{\mathrm{exp}} \approx 50$ to 100 nm [458–460] of triple junction microcracks in plastically deformed metals. Microcracks with the same characteristic length have been observed at triple junctions of interphase boundaries in the composite ceramics ZrO_2–Al_2O_3 [461].

We now make an analogous calculation of the equilibrium length of a microcrack generated at one of the small-strength disclinations resulting from splitting of the initial junction disclination when the local amorphous phase is formed (Fig. 5.3b). Consider an ensemble of N identical straight wedge disclinations (each with strength $-\omega = -\Omega/N$) located within an isotropic elastic cylinder with radius R. The position of the ith disclination is characterized by the radius vector R_i and azimuthal angle φ_i.

For definiteness, we choose coordinates in such a way as to have the disclination with number $i = 1$ and coordinates R_1 and φ_1 located on the x axis, that is, φ_1 is chosen to be zero (Fig. 5.3b). The stress function for such a disclination shifted from the cylinder axis has been obtained by Romanov as follows [54] (see also [56, 263]):

$$\chi_1 = -\frac{D\omega}{4}\left\{\left[(x-R_1)^2+y^2\right]\ln\frac{R^2\left[(x-R_1)^2+y^2\right]}{(xR_1-R^2)^2+y^2R_1^2}+\frac{r^2(R_1^2-R^2)}{R^2}\right\}. \tag{5.15}$$

Rotating the reference frame (x, y) through an angle φ_i and using cylindrical coordinates (r, θ), we transform the above stress function into the following stress function for the i th disclination:

$$\chi_i = -\frac{D\omega}{4}\left[(r^2+R_i^2-2rR_i\alpha_i)\ln\frac{R^2(r^2+R_i^2-2rR_i\alpha_i)}{R^4+r^2R_i^2-2rR^2R_i\alpha_i}+\frac{r^2(R_i^2-R^2)}{R^2}\right], \tag{5.16}$$

using the parameter $\alpha_i = \cos(\theta - \varphi_i)$. The tangential component $\sigma_{\theta\theta}^{(i)}$ of the stress field of the i th disclination is obtained from the stress function (5.16)

$$\begin{aligned}\sigma_{\theta\theta}^{(i)} = \frac{\partial^2\chi_i}{\partial r^2} = -\frac{D\omega}{2}\Bigg[&\ln\frac{R^2(r^2+R_i^2-2rR_i\alpha_i)}{R^4+r^2R_i^2-2rR^2R_i\alpha_i}+\frac{2(r-R_i\alpha_i)^2}{r^2+R_i^2-2rR_i\alpha_i}\\ &+\frac{2R^2R_i^2(R^2-R_i^2)(R^2-r^2)(1-\alpha_i^2)}{(R^4+r^2R_i^2-2rR^2R_i\alpha_i)^2}\\ &+\frac{2R_i(rR_i-R^2\alpha_i)(rR_i^2+R^2R_i\alpha_i-2rR^2)}{R^2(R^4+r^2R_i^2-2rR^2R_i\alpha_i)}\\ &-\frac{R_i^2(R^2-R_i^2)(R^2-r^2)}{R^2(R^4+r^2R_i^2-2rR^2R_i\alpha_i)}\Bigg].\end{aligned} \tag{5.17}$$

Equation (5.17) for $\sigma_{\theta\theta}^{(i)}$ is rather cumbersome. However, it simplifies in the situation $R_i = R_0 \ll R$, $r \ll R$, which is of interest for our analysis of an amorphized triple junction of grain boundaries (Fig. 5.3b). In this situation, we obtain from (5.17) the following approximate formula for $\sigma_{\theta\theta}^{(i)}$:

$$\sigma_{\theta\theta}^{(i)}(r,\ R_0 \ll R) \approx D\omega\left[-1+\ln\frac{R}{r_i}+(1-\alpha_i^2)\frac{R_0^2}{r_i^2}\right], \tag{5.18}$$

where $r_i^2 = r^2 + R_0^2 - 2rR_0\alpha_i$. Then the sum stress field $\sigma_{\theta\theta}^{(N)}$ of all disclinations can be written as the superposition of fields $\sigma_{\theta\theta}^{(i)}$, viz.,

$$\sigma_{\theta\theta}^{(N)}(r,\ R_0 \ll R) \approx D\Omega\left\{-1+\frac{1}{N}\sum_{i=1}^{N}\left[\ln\frac{R}{r_i}+(1-\alpha_i^2)\frac{R_0^2}{r_i^2}\right]\right\}. \tag{5.19}$$

To obtain the equilibrium length of a microcrack, we once again need the mean weighted stress $\bar{\sigma}_{\theta\theta}^{(N)}$. A detailed calculation of this sum stress is presented in the Appendix to [206], where it is shown that one has the following effective approximate formula for $\bar{\sigma}_{\theta\theta}^{(N)}$:

$$\bar{\sigma}_{\theta\theta}^{(N)} \approx D\Omega\left(\ln\frac{R}{R_0}-\frac{1}{2}\right) , \tag{5.20}$$

which will be used in our further estimates.

Let us examine the equilibrium length l_N of a microcrack generated in the field (5.20) at the i th disclination and located along the grain boundary with $\theta = 0$ (Fig. 5.3b). It should be noted that we are not interested in the absolute value of l_N. In order to characterize how amorphization influences microcrack generation, we need only the ratio l_N/l_1. Amorphization hampers microcrack generation at a triple junction, if $l_N/l_1 > 1$. In contrast, if $l_N/l_1 < 1$, amorphization stimulates microcrack generation. From (5.12), we have

$$l_N/l_1 = [\bar{\sigma}_{\theta\theta}^{(1)}/\bar{\sigma}_{\theta\theta}^{(N)}]^2 . \tag{5.21}$$

Substituting (5.13) and (5.20) into (5.21) results in

$$l_N/l_1 = \left[\frac{\ln(R/l_1)+2\ln 2-2}{\ln(R/R_0)-1/2}\right]^2 . \tag{5.22}$$

Equation (5.22) indicates that, when the radius R_0 of the amorphous embryo increases, so does the ratio l_N/l_1. Hence, for values $R = 10^{-4}$ m and $l_1 \approx 67$ nm (see above), the ratio l_N/l_1 grows from 2.7 to 13.8 when the amorphous-phase radius R_0 increases from 1 to 10 μm. This means that the growth of the amorphous phase at a triple junction of grain boundaries decreases the probability of microcrack generation at this triple junction.

This conclusion is in good accordance with experimental HREM images of triple junctions of interphase boundaries in the composite ceramics ZrO_2–Al_2O_3 studied by Pippel and Woltersdorf [461]. They have observed two modes of relaxation of high elastic stresses near triple junctions: the generation of microcracks propagating along the ZrO_2/Al_2O_3 interfaces, and the amorphization of material around the triple junctions. It is worth noting that the general view and character of the observed microcrack arrangement can easily be described in terms of triple junction disclinations. Indeed, in [461], the ZrO_2/Al_2O_3 interphase boundary is shown. This serves as a barrier for twinned ZrO_2 lamellae of martensite type. The edges of the twin boundaries (between ZrO_2 lamellae) at the ZrO_2/Al_2O_3 interface have to contain triple junction disclinations of alternative signs. At negative disclinations, in the elastically stretched regions, microcracks can be generated. However, they must be closed near positive disclinations, in the elastically compressed area. This arrangement of microcracks is demonstrated in [461].

The results presented [206] allow us to conclude that triple junction disclinations in a plastically deformed polycrystalline solid can serve as defects which stimulate either microcrack generation or amorphization (hampering microcrack generation), depending on the features of the solid and the deformation regime. At low levels of plastic deformation (when the densities of defects are small, $\varepsilon_{\mathrm{a-c}}$ is large), disclinations play the role of defects initiating microcrack generation. At high levels of plastic deformation (when the densities of defects are high, $\varepsilon_{\mathrm{a-c}}$ is small), disclinations play the role of defects initiating amorphization which impedes microcrack generation. Thus triple junction disclinations play a double role: they decrease plasticity at low plastic strains and increase plasticity at high plastic strains.

5.2 Splitting of Triple Junction and Grain Boundary Disclinations into Small-Strength Grain Boundary Disclinations

The theoretical approach developed in [206] to analyse the splitting of triple junction disclinations and the results discussed in the previous section have been used as a basis for further investigation of the behavior of triple junction [426, 427] and grain boundary [425–427] disclinations. The splitting of an initial grain boundary (or triple junction) disclination into smaller-strength grain boundary disclinations distributed along one or simultaneously along three grain boundaries has been considered in detail. Below, we briefly discuss only the main results of this work.

Calculations of strain energies for the split disclination configurations have shown that the split arrangements are always more energetically favorable than the initial disclinations. (We have not taken into account the energy of stacking faults which can form between the resulting disclinations. With this energy, the problem amounts to searching for the critical strength of the initial disclination, in a way similar to that described in Sect. 5.1.2.) However, the choice of the most favorable split configuration depends strongly on the conditions of splitting. For example, when the splitting of a grain boundary disclination occurs along the boundary segment, which is so long that it can be treated as an infinite straight line, the most favored configuration is the arrow of disclinations with minimum strengths and maximum interdisclination spacing. If the region of splitting is a boundary segment with a fixed length, then the most favored configuration is the pair of half-strength disclinations which are localized at the edges of the boundary segment. The first situation is typical for conventional polycrystals, whereas the second occurs in NCMs, where triple junctions of grain boundaries can play the role of these edge points.

It is also of interest to compare the gains in strain energy for those split configurations where the resulting disclinations are periodically distributed either along three equal boundary segments with a common triple junction and equal angles between the segments or along a circle with its center at the triple junction (see Sect. 5.1) and radius equal to the segment length.

When the initial triple junction disclination splits into three new disclinations with equal strengths and localized either at the three edges of the boundary segments or along the circle, then the resulting disclination configurations are equivalent and give equal gains in total strain energy. Further splitting into smaller-strength disclinations, which are periodically distributed along either the boundary segments or the circle, gives an energy gain which either significantly decreases with an increasing number N of new disclinations or is practically independent of N, respectively. In the first case, for example, the energy gain at $N = 18$ is approximately a factor of two smaller than at $N = 3$.

Calculations of the equilibrium lengths of microcracks generated at the split disclination configurations have shown that the splitting increases the equilibrium length of a microcrack, thus diminishing the probability of its generation, as is the case with amorphization of a triple junction. However, this increase in the equilibrium lengths is not significant: the ratio l_N/l_1 achieves the value ~ 1.5, in contrast to $\sim$ 3–14 which is typical for amorphization (see Sect. 5.1.3). This is because, in the situation with amorphization, the microcrack is opened relatively far from most of the new disclinations, while in the other models, it crosses the lines either of all new disclinations or of one third of them.

In comparing the results of the present and previous sections, one can conclude that the splitting of triple junction disclinations (if it is possible) accompanied by local amorphization of the triple junctions is a far more effective channel for elastic stress relaxation than the other types of splitting considered. Thus, local amorphization can become (under definite conditions) a real alternative to microcrack generation.

5.3 Nanocrack Generation at a Wedge Disclination Near a Triple Junction of Grain Boundaries

Failure processes in NCMs are strongly influenced by grain boundaries whose volume fraction ranges from 10 to 50% in such materials (see Sect. 1.2). More precisely, cracks in NCMs are commonly intercrystalline: they nucleate and grow along grain boundaries [462]. In this context, with the very high volume fraction of triple junctions in NCMs, the behavior of intercrystalline cracks at triple junctions where grain boundary planes change their orientation is of special interest. The main aim of this section is to consider a theoretical model [444] that describes nucleation and growth of nanoscale cracks, hereafter referred to as nanocracks, along grain boundaries, taking into account effects of the triple junction geometry.

It seems evident that the first nanocracks must be generated inhomogeneously, at defects and stress concentrators. Triple junction and grain boundary disclinations can be considered as the most probable sites for such generation. The key statements of the model [444] are as follows:

- Nanocracks grow along grain boundaries and change the orientation of their growth at triple junctions.
- Nanocracks nucleate and grow in the stress fields of disclinations.
- Nucleation and growth of nanocracks are described using the configurational force approach (see Sect. 5.1.3).

Hence, the nucleation and growth of a nanocrack is assumed to be driven by a release of the stresses created by a disclination located at either a grain boundary or a triple junction (Fig. 5.4). In the framework of our model [444], we consider a negative wedge disclination of strength $-\omega$, located in an elastically isotropic solid cylinder with radius R and infinite length. The disclination line is parallel to the central axis of the cylinder and situated at a distance R_1 from it. The nanocrack is nucleated at the disclination. It grows along the grain boundary plane AO and reaches the triple junction line that coincides with the central axis of the cylinder. The nanocrack changes the direction of its growth at the triple junction. It starts to grow along one of the adjacent grain boundaries. In this situation, the nanocrack can be divided into the first and second segments, oriented along the two grain boundaries AO and OB, respectively. In the framework of this model, during growth of the first nanocrack segment towards the triple junction, the shear components of the disclination stress tensor are equal to zero. That is, the first nanocrack segment grows as a mode I crack under the action of the tensile stress $\sigma_{\theta\theta}$. The second nanocrack segment grows by both I (normal) and II (plane shear) modes.

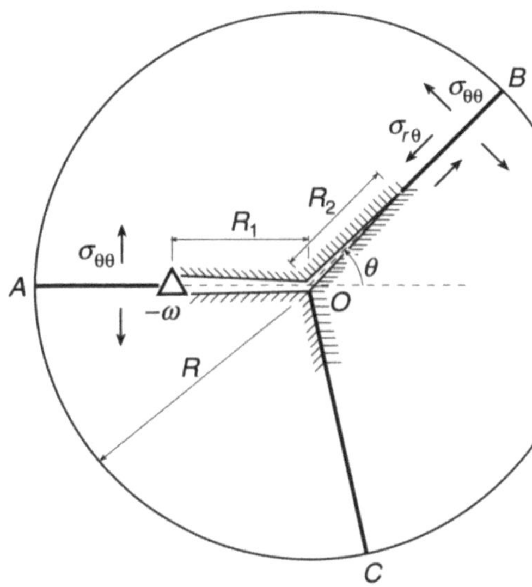

Fig. 5.4. Generation of a curved nanocrack at a wedge disclination near a triple junction (O) of grain boundaries AO, OB and OC, in a free-surface cylinder

It is intuitively clear that the second segment of the intercrystalline nanocrack grows along the OB boundary plane which makes a low or intermediate slope angle θ with the AO boundary plane, in contrast to the OC boundary plane, highly misoriented relative to the AO plane. In the situation discussed, the triple junction geometry is characterized by the angle θ, and the triple junction effect on the nanocrack growth can be effectively described by relationships between characteristics of the nanocrack and the angle θ. For definiteness, we will focus our analysis on the relationship between θ and the equilibrium lengths R_1 and R_2 of the first and second nanocrack segments, respectively, which characterize the probability of the energetically favorable formation of the nanocrack curving at the triple junction (Fig. 5.4).

Thus, the first segment of the nanocrack grows under the action of the disclination stress component $\sigma_{\theta\theta}(r, \theta = \pi)$, while the second one grows under both the components $\sigma_{\theta\theta}(r, \theta)$ and $\sigma_{r\theta}(r, \theta)$. These stress components can be obtained with the help of the stress function χ given by (5.16), where one should replace R_i by R_1 and take $\alpha_i = -\alpha = -\cos\theta$. As a result, the normal component $\sigma_{\theta\theta}$ is determined by (5.17) with the same replacements. The shear stress component $\sigma_{r\theta}$ then reads [444]

$$\begin{aligned}\sigma_{r\theta} &= -\frac{\partial}{\partial r}\left(\frac{1}{r}\frac{\partial\chi}{\partial\theta}\right)\\ &= -D\omega R_1 \sin\theta \left[\frac{r + R_1\alpha}{r^2 + R_1^2 + 2rR_1\alpha} - \frac{R^2(r + R_1\alpha)}{R^4 + r^2R_1^2 + 2rR_1R^2\alpha}\right.\\ &\quad \left. - \frac{R_1(R^2 - R_1^2)(R^2 - r^2)(rR_1 + R^2\alpha)}{(R^4 + r^2R_1^2 + 2rR_1R^2\alpha)^2}\right] . \qquad (5.23)\end{aligned}$$

For definiteness and simplicity, we focus our attention on the limiting case $r \ll R$ and $R_1 \ll R$. This corresponds to the initial stage of growth of the second nanocrack segment, the stage at which the nanocrack becomes curved due to the triple junction effect. In the case discussed, the stress components are

$$\sigma_{\theta\theta}\Big|_{r,R_1 \ll R} \approx D\omega\left[\ln\frac{R}{r_1} + (1 - \cos^2\theta)\frac{R_1^2}{r_1^2} - 1\right] , \qquad (5.24)$$

$$\sigma_{r\theta}\Big|_{r,R_1 \ll R} \approx -D\omega R_1 \sin\theta \frac{r + R_1\cos\theta}{r_1^2} , \qquad (5.25)$$

where $r_1^2 = r^2 + R_1^2 + 2rR_1\cos\theta$.

Equations (5.24) and (5.25) describe the stress field in which the nucleation and growth of the nanocrack occur (Fig. 5.4). Let us calculate to the first approximation the equilibrium length $L = R_1 + R_2$ of the nanocrack, using the configurational force method [457] (see also Sect. 5.1.3). The configurational force F is defined as the elastic energy released when the crack moves over a unit distance. In the situation with the plane strained state examined here, F can be written in the general form [457]

$$F = \frac{\pi(1-\nu)L}{4G}\left(\bar{\sigma}_{\theta\theta}^2 + \bar{\sigma}_{r\theta}^2\right) , \tag{5.26}$$

where $\bar{\sigma}_{\theta\theta}$ and $\bar{\sigma}_{r\theta}$ are the mean weighted values of the stress tensor components $\sigma_{\theta\theta}$ and $\sigma_{r\theta}$, respectively. With the stress tensor components from (5.24) and (5.25), we find:

$$\bar{\sigma}_{\theta\theta} = \frac{2}{\pi L}\left[\int_0^{R_1} \sigma_{\theta\theta}(r, \theta = \pi)\sqrt{\frac{r}{R_1 - r}}\mathrm{d}r + \int_0^{R_2} \sigma_{\theta\theta}(r, \theta)\sqrt{\frac{r}{R_2 - r}}\mathrm{d}r\right] , \tag{5.27}$$

$$\bar{\sigma}_{r\theta} = \frac{2}{\pi L}\int_0^{R_2} \sigma_{r\theta}(r, \theta)\sqrt{\frac{r}{R_2 - r}}\mathrm{d}r . \tag{5.28}$$

The equilibrium length L of the nanocrack is derived from the balance $F = 2\gamma$ between the release F of elastic energy and the formation of two new nanocrack surfaces characterized by the surface energy density γ per unit area. With this balance and (5.26), we have

$$\frac{8G\gamma}{\pi(1-\nu)L} = \bar{\sigma}_{\theta\theta}^2 + \bar{\sigma}_{r\theta}^2 . \tag{5.29}$$

Thus, in order to calculate L, it is necessary to calculate the integrals in (5.27) and (5.28) and then solve (5.29).

Calculating the integrals using the tables given in [463], we find

$$\bar{\sigma}_{ij} = \frac{D\omega}{L} g_{ij}(R_1, R_2, \theta) , \tag{5.30}$$

$$g_{\theta\theta} = R_1 \ln\frac{4R}{R_1} + R_2\left[\ln\frac{R}{R_1} - 1 - \frac{\psi(R_1/R_2, \theta)}{\pi} + \frac{2R_1^2 \sin^2\theta}{d\sqrt{2R_1(d+l)}}\right] , \tag{5.31}$$

$$g_{r\theta} = R_1\left[2\frac{R_1^3}{R_2^3}\sqrt{\frac{R_1}{d}}\sin(\theta - \phi/2) + \frac{R_2}{d}\sqrt{\frac{R_1}{2(d+l)}}\sin(2\theta) - 2\sin\theta\right] . \tag{5.32}$$

Here

$$\psi(R_1/R_2, \theta) = \int_0^1 \sqrt{\frac{x}{1-x}}\ln\left(x^2 + 2x\frac{R_1}{R_2}\cos\theta + \frac{R_1^2}{R_2^2}\right)\mathrm{d}x , \tag{5.33}$$

and the geometric interpretation of d ($d^2 = R_1^2 + R_2^2 + 2R_1R_2\cos\theta$), $l = R_1 + R_2\cos\theta$ and $\phi = \arctan[R_2\sin\theta/(R_1 + R_2\cos\theta)]$ is evident from Fig. 5.5.

With the standard estimate $\gamma \approx Gb/8$, where b is the interatomic distance, (5.29) can be rewritten in the form

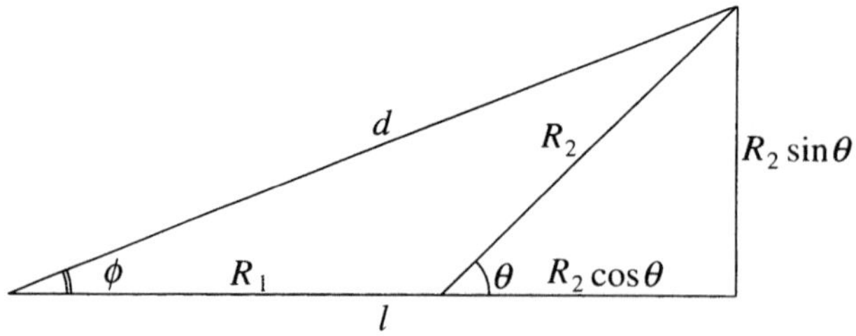

Fig. 5.5. Geometric relationships between parameters in the problem of the curved nanocrack

$$\frac{4\pi(1-\nu)Lb}{\omega^2} = g_{\theta\theta}^2 + g_{r\theta}^2 \,. \tag{5.34}$$

In the general situation, this transcendental equation can only be solved numerically. Here we consider the limiting case $R_2 \ll R_1$, which is interesting for a description of the real situation where a nanocrack 'chooses' a new direction of growth at a triple junction of grain boundaries. In this case, (5.34) has an approximate analytical solution. More precisely, for $R_2 \ll R_1$, the integral (5.33) can be approximated by

$$\psi(R_1/R_2 \gg 1, \theta) \approx \pi \left(\ln \frac{R_1}{R_2} + \frac{3}{4}\frac{R_2}{R_1}\cos\theta + \frac{5}{16}\frac{R_2^2}{R_1^2} \right) , \tag{5.35}$$

and functions (5.31) and (5.32), with $d \approx l \approx R_1$ and $\phi \approx 0$, become

$$g_{\theta\theta}(R_2 \ll R_1, \theta) \approx R_1 \left[\ln \frac{4R}{R_1} + \frac{R_2}{R_1}\left(\ln \frac{R}{R_1} - \cos^2\theta \right) \right] . \tag{5.36}$$

$$g_{r\theta}(R_2 \ll R_1, \theta) \approx 2R_1 \sin\theta \left(\frac{R_1^3}{R_2^3} - 1 + \frac{R_2}{R_1}\cos\theta \right) . \tag{5.37}$$

Taking the second power of (5.36) and (5.37) and keeping only the key terms yields

$$g_{\theta\theta}^2 \approx R_1^2 \ln \frac{4R}{R_1} \left[\ln \frac{4R}{R_1} + 2\frac{R_2}{R_1}\left(\ln \frac{R}{R_1} - \cos^2\theta \right) \right] \approx R_1^2 \ln^2 \frac{4R}{R_1} , \tag{5.38}$$

$$g_{r\theta}^2 \approx 4R_1^2 \sin^2\theta \frac{R_1^3 - R_2^3}{R_2^3} \left(\frac{R_1^3 - R_2^3}{R_2^3} + \frac{2R_2}{R_1}\cos\theta \right) \approx 4R_1^2 \sin^2\theta \frac{R_1^6}{R_2^6} \,. \tag{5.39}$$

Substituting (5.38) and (5.39) into (5.34) and using the approximation $L = R_1 + R_2 \approx R_1$, we find that

$$\sin\theta \approx \frac{R_2^3}{R_1^3} \sqrt{\frac{\pi(1-\nu)b}{\omega^2 R_1} - \frac{1}{4}\ln^2 \frac{4R}{R_1}} \,. \tag{5.40}$$

When the term on the right-hand side of (5.40) is small (it is commonly valid at realistic values of parameters of the system), this formula reduces to

$$\frac{R_2}{R_1} \approx \theta^{1/3} \left[\frac{\pi(1-\nu)b}{\omega^2 R_1} - \frac{1}{4} \ln^2 \frac{4R}{R_1} \right]^{-1/6} . \tag{5.41}$$

Consequently, the formation of a curved intercrystalline nanocrack (Fig. 5.4) in an NCM is characterized by the equilibrium length $L = R_1 + R_2$. Segments of the nanocrack have scales R_1 and R_2 related by (5.41) with the same angle θ characterizing the triple junction geometry. At the initial stage of growth of the second nanocrack segment, its length R_2 is proportional to $\theta^{1/3}$, in which case the formation of curved nanocracks is hampered at triple junctions with large values of θ. The dependence (5.41) is rather strong. Here θ is given in radians and is therefore smaller than 1 for most cases of interest ($\theta \leq 57°$). In the range $\theta \leq 34°$, the function $\theta^{1/3}$ goes even higher than the exponential function, $(e^{\theta} - 1)$, and in the range $\theta \leq 57°$, it gives higher values than the function θ^n, where $n \geq 1$. Moreover, the first derivative of $\theta^{1/3}$ is higher than the exponential function e^{θ} in the range $\theta \leq 10°$. This means that changes in θ must have a significant effect on the strength of a nanomaterial.

In the context of the model [444], devoted mainly to the effect of the triple junction geometry, the disclination strength ω is used as a fixed parameter. Its strong influence on nanocrack generation is well known (see Sect. 5.1.3). For the simple case of a straight crack generated at a straight disclination, we have shown [206] the following relation between the equilibrium length L of the crack and ω: $L \sim \omega^{-2}$. In the present section, a similar but more cumbersome relationship follows directly from (5.34) with (5.31)–(5.33) taken into account. It seems to be impossible to predict the exact value of ω at the present level of theoretical development, but there can be no doubt that it greatly depends on the technology of production and treatment of a nanocrystalline sample.

In as-prepared samples, produced by severe plastic deformation, ω may reach a value ≈ 0.1 (5°–7°). In well-annealed samples, it must be several times smaller. To obtain these estimates, one can use typical magnitudes of Burgers vectors b and spacing l for grain boundary dislocations. In the first case, $b \approx 0.1$ nm and $l \approx 1$ nm, which give $\omega \approx b/l \approx 0.1$. In the second case, l must increase and ω must therefore decrease. However, the equilibrium length of a nanocrack depends, not only on ω, but also on the difference between the surface energy of the crack borders (γ_s) and the effective surface energy of the grain boundary (γ_{gb}), $\gamma = \gamma_s - \gamma_{gb}$, as well as on the value of R (see Fig. 5.4), which plays the role of a screening radius for the disclination stress field. For example, for $\omega = 0.1$, let us take $R \approx 1000$ nm and $\gamma = \gamma_s - \gamma_{gb} = \gamma_s/n$, where n characterizes the strength of the grain boundary. ($n = 1$ means that the grain boundary has the same strength as the intragrain material, while $n > 1$ shows how many times the boundary strength is weaker.) For the typical estimate $\gamma_s = Gb/8$ and $n = 1$, 2, and 5, this set of parameters gives $L \approx 66$ nm, 8 nm, and 1.6 nm, respectively. Thus, the weaker grain boundary is the most probable site for nanocrack generation. For a real as-prepared nanomaterial, the estimate $n \approx 2$ seems to be quite realistic. We may therefore expect $L \approx 8$ nm and the experimentally observed nanocracks must be larger than 8 nm. (Smaller cracks must be closed.) Of course, for a specific material, we have to know the characteristic values for all these parameters (ω, R, γ_{gb}) in order to predict the value of L.

Coming back to the results of the present section, let us discuss briefly the influence of thermal annealing on the conditions of nanocrack generation. First, annealing reduces ω because the structure of grain boundaries becomes a more equilibrium structure. Second, it transforms the triple junctions of grain boundaries into those with 'equilibrium' dihedral angles of order 120° between adjacent grain boundaries. Both these factors must increase the nanocrack equilibrium lengths, thus diminishing the probability of their generation. We can conclude that, to get a stronger nanomaterial, we have to design highly equilibrium low-energy grain boundaries without disclinations and with equilibrium triple junctions that have dihedral angles of order 120°. In terms of considerations given here, the heat treatment transforms, in particular, triple junctions with low values of θ into those with $\theta \approx 60°$, thus strengthening a heat-treated NCM. This conclusion is in agreement with experimental data [321, 464] on the heat-treatment-induced strengthening of NCMs. In general, different relaxation mechanisms – deformation mechanisms and nanocrack formation – are realized at different triple junctions, depending on their geometric parameters. In this context, the results of the model [444] can also be used to describe competition between the relaxation mechanisms that cause the unique deformation behavior of NCMs.

The results represented in Chap. 5 allow us to draw the following conclusions:

- Splitting of triple junction disclinations in nano- or polycrystalline solids can be accompanied by local solid-state amorphization of the triple junctions and formation of ensembles of smaller-strength disclinations at the boundaries of the amorphous embryos. Such splitting can take place when the strength of the initial triple junction disclination is larger than a critical value. In conventional polycrystals, these splitting processes are possible in the late stages of plastic deformation, when an increase in the density of defects leads to a significant decrease in the difference between the energy densities of the amorphous and crystalline states.
- The splitting of a triple junction disclination strongly increases the equilibrium length of the microcrack (i.e., decreases the probability of its appearance) which is generated at the triple junction. Local amorphization of a triple junction can be treated as an alternative to microcrack generation at the triple junction. Triple junction disclinations in plastically deformed nano- and polycrystalline solids can stimulate either microcrack generation or amorphization at the triple junctions, depending on the features of the solid and the deformation regime. At low levels

of plastic deformation, disclinations initiate microcrack generation. At high levels of plastic deformation, they initiate local amorphization of the triple junction, which impedes microcrack generation. Thus triple junction disclinations play a double role: they decrease plasticity of polycrystals at low plastic strains and increase their plasticity at high plastic strains.

- Splitting of grain boundary disclinations also diminishes the strain energy of a nano- or polycrystalline solid and decreases the probability of nano- or microcrack generation at grain boundaries.
- Triple junctions of grain boundaries serve as effective obstacles for nanocrack growth along grain boundaries in NCMs. The larger the angle between grain boundaries adjacent to a triple junction, the larger the equilibrium length of the curved nanocrack and, therefore, the smaller the probability of its generation.

6 Conclusion

We have considered theoretical models of plastic flow in NCMs, paying special attention to the abnormal Hall–Petch effect, localization of deformation, crossover to rotational modes of plasticity, amorphization and generation of micro- and nanocracks. On the basis of the results presented, one can draw the following general conclusions:

- Different theoretical models give different explanations of the abnormal Hall–Petch relationship, and most of them account well for the corresponding experimental data. However, it is extremely difficult to identify the deformation mechanism(s) in NCMs experimentally due to their very complicated nanoscale structure and the transformations it undergoes at various length scales during plastic deformation. In addition, the deformation mechanisms may be different in different NCMs or even in the same material under different conditions of loading (e.g., temperature, strain rate). The main mechanisms, which determine the first stages of plastic deformation in NCMs, are gliding of partial or perfect lattice dislocations (for relatively coarse-grained NCMs, $d \geq 40$ nm) and grain boundary sliding and diffusional plasticity (for relatively fine-grained NCMs, $d \leq 40$ nm). Competition between these mechanisms, depending on structural and material properties of NCMs, as well as on conditions of external loading, leads to the abnormal Hall–Petch relationship.
- The regimes of homogeneous and inhomogeneous high-strain plastic deformation in NCMs are effectively described with the help of the concept of cooperative grain boundary sliding and the model of cellular dislocations whose elastic fields can be modelled by nonsingular solutions of gradient elasticity (both the nonlocal and gauge theories give the same solutions), while their kinetics can be studied using the presented solution of the system of evolutionary equations. The degree of inhomogeneity of plastic deformation is determined by the intensity of the accommodation processes which accompany grain boundary sliding: low intensity results in the homogeneous regime of plastic flow and high intensity in the inhomogeneous regime.
- Rotational plastic deformation occurs in NCMs through generation and development of specific (rotational) structures of lattice and/or grain boundary dislocations which are effectively described as partial discli-

nations. The rotational structures are generated in NCMs at various imperfections of grain boundaries (kinks, double and triple junctions), where the misorientation angle changes sharply. The paths for the development of rotational plasticity depend on the dominant mechanisms of translational plasticity. Dominance of the lattice gliding mechanism can lead to the appearance of misorientation bands (or other disclination structures which are typical for conventional metals and alloys but have not been considered here) inside the grains or to motion of grain boundary disclinations by issuing lattice dislocations. Dominance of grain boundaries can result in the formation and motion of grain boundary disclinations through climb of grain boundary dislocations. In both these situations, the motion of grain boundary disclinations along their grain boundaries is accompanied by a change in the grain misorientations under external loading and is capable of producing the corresponding rotation of the grain crystalline lattice as a whole.

- In the late stages of plastic deformation in nano- and polycrystalline solids, local solid-state amorphization or generation of nano- and microcracks, respectively, can occur at triple junction and grain boundary disclinations. At low levels of plastic deformation, disclinations initiate microcrack generation. At high levels of plastic deformation, they initiate local amorphization of the triple junctions, which impedes microcrack generation. This means that such disclinations play a double role: they decrease plasticity at low plastic strains and increase plasticity at high plastic strains. Triple junctions of grain boundaries serve as effective obstacles for nanocrack growth along grain boundaries in NCMs. The larger the angle between grain boundaries adjacent to a triple junction, the larger the equilibrium length of the curved nanocrack and hence the smaller the probability of its generation.

In these circumstances, further theoretical and experimental investigations in this area are highly desirable in order to understand the fundamentals of the outstanding deformation behavior of NCMs and develop new technologies exploiting their unique mechanical properties.

To conclude this book, let us outline the key points which are interesting for future theoretical studies of plastic deformation processes and associated phenomena in NCMs:

- Description of new deformation mechanisms in NCMs.
- Analysis of the role of triple junctions and quadruple nodes of grain boundaries in deformation processes in NCMs.
- Development of a generalized model which would describe the combined action of different deformation mechanisms (such as lattice dislocation motion in grain interiors, grain boundary sliding, diffusion plasticity mechanisms associated with grain boundary diffusion and triple junction diffusion) whose contribution to the plastic flow in NCMs is dependent on the material characteristics, grain size distribution and loading conditions.

- Investigation of the specific features of plastic flow in nanocrystalline films and coatings, where the residual stresses are capable of essentially influencing grain boundary structures and deformation processes.
- Study of the specific features of plastic flow in nanocomposite materials where the composite structure and interphase boundaries should play a leading role in the development of plastic deformation.
- Description of the influence of plastic deformation on structural stability (against grain growth) in NCMs.

References

1. M.C. Roco, R.S. Williams, P. Alivisatos (Eds.): *Nanotechnology Research Directions: IWGN Workshop Report. Vision for Nanotechnology R&D in the Next Decade* (Kluwer Academic Publishers, Dordrecht Boston London 2000)
2. M.C. Roco: National nanotechnology investment in the FY 2004 budget request (`http://www.nano.gov`)
3. *EU Six Framework Programme* (`http://fp6.cordis.lu`)
4. M.C. Roco, W.S. Bainbridge (Eds.): *Societal Implications of Nanoscience and Nanotechnology* (Kluwer Academic Publishers, Dordrecht 2001)
5. E. Cartlidge: Nanotechnology: An industry still on the starting blocks, Physics World **14**, No. 4, 42 (2001)
6. S. Komarneni, R.A. Vaia, G.Q. Lu, J.-I. Matsushita, J.C. Parker (Eds.): *Nanophase and Nanocomposite Materials IV. MRS Symp. Proc.*, Vol. 703 (MRS, Warrendale 2003)
7. C.C. Berndt, T. Fischer, I.A. Ovid'ko, G. Skandan, T. Tsakalakos (Eds.): *Nanomaterials for Structural Applications. MRS Symp. Proc.*, Vol. 740 (MRS, Warrendale 2003)
8. M.Yu. Gutkin, I.A. Ovid'ko: *Defects and Mechanisms of Plasticity in Nanostructured and Non-Crystalline Materials* (Yanus, Saint Petersburg, 2001) in Russian
9. R. Birringer, H. Gleiter: Nanocrystalline materials. In: *Advances in Materials Science, Encyclopedia of Materials Science and Engineering*, Vol. 1, ed. by R.W. Cahn (Pergamon Press, Oxford 1988) pp. 339–349
10. H. Gleiter: Nanocrystalline materials, Progr. Mater. Sci. **33**, 223–315 (1989)
11. H. Gleiter: Nanostrukturierte materialien, Phys. **47**, No. 8, 753–759 (1991)
12. H. Gleiter: Materials with ultrafine microstructures: Retrospective and perspective, Nanostruct. Mater. **1**, No. 1, 1–19 (1992)
13. U. Erb, A.M. El-Sherik, G. Palumbo, K.T. Aust: Synthesis, structure and properties of electroplated nanocrystalline materials, Nanostruct. Mater. **2**, No. 4, 383–390 (1993)
14. R.W. Siegel: Synthesis and properties of nanophase materials, Mater. Sci. Eng. A **168**, No. 2, 189–197 (1993)
15. V.G. Gryaznov, L.I. Trusov: Size effect in micromechanics of nanocrystals, Progress in Material Science **37**, No. 4, 290–400 (1993)
16. R.W. Siegel: Nanostructured materials – Mind over matter, Nanostruct. Mater. **4**, No. 1, 121–138 (1994)
17. H.-E. Schaefer, R. Würschum, H. Gleiter, T. Tsakalakos (Eds.): *Proceedings of the Second International Conference on Nanostructured Materials (Stuttgart, Germany, October 3–7, 1994)*, Nanostruct. Mater. **6**, Nos. 5–8, 533–1026 (1995)

18. A.S. Edelstain, R.C. Cammarata (Eds.): *Nanomaterials: Synthesis, Properties and Applications* (Institute of Physics Publ., Bristol Philadelphia 1996)
19. N.I. Noskova: Properties of nanocrystalline multiphase alloys. In: *Contemporary Problems of Physics and Mechanics of Materials* (Saint Petersburg State University, Saint Petersburg 1997) pp. 333–349, in Russian
20. *Structure, Phase Transformations and Properties of Nanocrystalline Alloys* (Ural Department of Russian Academy of Sciences, Ekaterinburg 1997), in Russian
21. R.W. Siegel, E. Hwu, M.C. Roco (Eds.): *R&D Status and Trends in Nanoparticles, Nanostructured Materials, and Nanodevices in the United States* (International Technology Research Institute, Baltimore 1997)
22. G.-M. Chow, N.I. Noskova (Eds.): *Nanostructured Materials: Science and Technology* (Kluwer Academic Publishers, Dordrecht Boston London 1998)
23. A.I. Gusev: Effects of the nanocrystalline state in solids, Uspekhi Fizicheskikh Nauk. **168**, No. 1, 55–83 (1998) in Russian [English translation in: Physics-Uspekhi **41**, No. 1, 49–76 (1998)]
24. R.A. Andrievskii, A.M. Glezer: Size effects in nanocrystalline materials. I. Structure characteristics, thermodynamics, phase equilibria, and transport phenomena, Fizika Metallov i Metallovedenie **88**, No. 1, 50–73 (1999) in Russian [English translation in: Phys. Metals Metallogr. **88**, No. 1, 45–66 (1999)]
25. R.A. Andrievskii, A.M. Glezer: Size effects in nanocrystalline materials. II. Mechanical and physical properties, Fizika Metallov i Metallovedenie **89**, No. 1, 91 (2000) in Russian [English translation in: Phys. Metals Metallogr. **89**, No. 1, 83 (2000)]
26. H.S. Nalwa (Ed.): *Handbook of Nanostructured Materials and Nanotechnology*, Vols. 1–5 (Academic Press, San Diego 1999)
27. Mechanical behavior of nanostructured materials, MRS Bulletin **24**, No. 2, 14–58 (1999)
28. R.Z. Valiev, I.V. Alexandrov: *Nanostructured Materials Obtained by Intensive Plastic Deformation* (Logos, Moscow 2000) in Russian
29. A.I. Gusev, A.A. Rempel: *Nanocrystalline Materials* (Nauka, Moscow 2000) in Russian
30. G.-M. Chow, I.A. Ovid'ko, T. Tsakalakos (Eds.): *Nanostructured Films and Coatings. NATO Science Series, High Technology*, Vol. 78 (Kluwer, Dordrecht 2000)
31. H. Gleiter: Nanostructured materials: Basic concepts and microstructure, Acta Mater. **48**, No. 1, 1–29 (2000)
32. R.A. Andrievskii: Nanostructured materials – State-of-art and applications, Perspektivnye Materialy **6**, 5–11 (2001) in Russian [English translation in: J. Advanced Mater.]
33. Yu.R. Kolobov, R.Z. Valiev, G.P. Grabovetskaya, A.P. Zhilyaev, E.F. Dudarev, K.V. Ivanov, M.B. Ivanov, O.A. Kashin, E.B. Naidenkin: *Grain-Boundary Diffusion and Properties of Nanostructured Materials* (Nauka, Novosibirsk 2001) in Russian
34. J. Eckert, A. Reger-Leonhard, B. Weiss, M. Heilmaier, L. Schultz: Bulk nanostructured multicomponent alloys, Adv. Eng. Mater. **3**, Nos. 1–2, 41–47 (2001)
35. D. Farkas, H. Kung, M. Mayo, J. Van Swygenhoven, J. Weertman (Eds.): *Structure and Mechanical Properties of Nanophase Materials – Theory and Computer Simulations vs. Experiment* (MRS, Warrendale 2001)

36. V.A. Zhorin, V.B. Fedorov, D.K. Khakimova, E.G. Galkina, E.V. Tatyanin, N.S. Enikolopyan: Formation of ultrafine structure in titanium nickelide at plastic flow under high pressure, Doklady AN SSSR **275**, No. 6, 1447–1449 (1984) in Russian [English translation in: Sov. Phys. Dokl.]
37. I.D. Morokhov, V.B. Fedorov, E.G. Galkina, E.B. Tatyanin: Investigation of heterogeneous structure and properties of titanium nickelide after pressure treatment. In: *Physics and Chemistry of Ultra-Dispersed Systems*, ed. by I.V. Tananaev (Nauka, Moscow 1987) pp. 157–160, in Russian
38. A.M. Glezer, B.V. Molotilov, V.P. Ovcharov, O.L. Utevskaya, Yu.Ye. Chicherin: Structure and mechanical properties of Fe–Cr–B alloys on transition from amorphous to crystalline state, Fizika Metallov i Metallovedenie **64**, No. 6, 1106–1109 (1987) in Russian [English translation in: Phys. Metals Metallogr.]
39. L.C. Wang, D. Feng, Q. Li, Y.L. He, Y.M. Chu: Microstructures and characteristics of nanosize crystalline silicon films, J. Phys.: Condens. Matter **4**, No. 40, L509–L512 (1992)
40. D.-H. Ko, R. Sinclair: Amorphous phase formation and initial interfacial reactions in the platinum/GaAs systems, J. Appl. Phys. **72**, No. 5, 2036–2042 (1992)
41. L.V. Iogansen: The possibility of resonance transmission of electrons in crystals through a system of barriers, Jurnal Eksperimental'noj i Teoreticheskoj Fiziki **45**, No. 2, 207–213 (1963) in Russian [English translation in: Sov. Phys. JETP-USSR **18**, No. 1, 146–150 (1964)]
42. L. Esaki, R. Tsu: Superlattice and negative differential conductivity in semiconductors, IBM J. Res. Developm. **14**, No. 1, 61–65 (1970)
43. J.W. Matthews, A.E. Blakeslee: Defects in epitaxial multilayers. 1. Misfit dislocations, J. Cryst. Growth **27**, No. 1, 118–125 (1974)
44. J.W. Matthews, A.E. Blakeslee: Defects in epitaxial multilayers. II. Dislocation pileups, threading dislocations, slip lines and cracks, J. Cryst. Growth **29**, No. 2, 273–280 (1975)
45. J.W. Matthews, A.E. Blakeslee: Defects in epitaxial multilayers. III. Preparation of almost perfect multilayers, J. Cryst. Growth **32**, No. 2, 265–273 (1976)
46. G.C. Osbourn, P.L. Gourley, I.J. Fritz, R.M. Biefeld, L.R. Dawson, T.E. Zipperian: Principles and applications of semiconductor strained-layer superlattices. In: *Semiconductors and Semimetals*, ed. by R.K. Willardson and A.C. Beer (Academic Press, New York 1987) Vol. 24, pp. 459–503
47. S. Sharan, J. Narayan, J.C.C. Fan: Dislocation density reduction in GaAs epilayers on Si using strained layer superlattices, J. Electron. Mater. **20**, No. 10, 779–784 (1991)
48. I. Sugiyama, A. Hobbs, T. Saito, O. Ueda, K. Shinohara, H. Takigawa: Dislocation reduction in HgCdTe epilayers on GaAs by using CdTe/CdSnTe strained-layer superlattices in CdTe layers, J. Cryst. Growth **117**, Nos. 1–4, 161–165 (1992)
49. Y.G. Gogotsi, R.A. Andrievski (Eds.): *Material Science of Carbides, Nitrides and Borides* (Kluwer Academic Publishers, Dordrecht Boston London 1999)
50. V.M. Fedosyuk: *Multilayered Magnetic Structures* (Belorussian State University, Minsk 2000) in Russian
51. D.V. Shtanskii, E.A. Levashov: Multicomponent nanostructured thin films: Problems and solutions, Izvestiya VUZov. Tsvetnaya Metallurgiya **3**, 52–62 (2001) in Russian [English translation in: Russian J. Non-Ferrous Metals]

52. R.A. Andrievski: Films as nanostructured materials with characteristic mechanical properties, Mater. Trans. **42**, No. 8, 1471–1473 (2001)
53. R.A. Andrievski, G.V. Kalinnikov: Physical-mechanical and physical-chemical properties of thin nanostructured boride/nitride films, Surf. Coat. Technology **142–144**, 573–578 (2001)
54. A.E. Romanov: Boundary-value problems in theory of elasticity for disclinations. In: *Experimental Investigation and Theoretical Description of Disclinations*, ed. by V.I. Vladimirov (Ioffe Physico-Technical Institute, Leningrad 1984) pp. 110–135, in Russian
55. V.G. Gryaznov, A.M. Kaprelov, A.E. Romanov: Pentagonal symmetry and disclinations in small particles. In: *Disclinations and Rotational Deformation of Solids*, ed. by V.I. Vladimirov (Ioffe Physico-Technical Institute, Leningrad 1988) pp. 47–83, in Russian
56. A.E. Romanov, V.I. Vladimirov: Disclinations in crystalline solids. In: *Dislocations in Solids*, Vol. 9, ed. by F.R.N. Nabarro (North-Holland, Amsterdam 1992) pp. 191–402
57. M.Yu. Gutkin, A.L. Kolesnikova, A.E. Romanov: Misfit dislocations and other defects in thin films, Mater. Sci. Eng. A **164**, Nos. 1–2, 433–437 (1993)
58. M.Yu. Gutkin: Models of defects and mechanisms of plastic deformation in heterogeneous media with meso- and nanostructure, Dr. Sci. Habil. Thesis, Institute of Problems of Mechanical Engineering, Saint Petersburg (1997) in Russian
59. A.E. Romanov: Micromechanics of defects in nanostructured materials. In: *Nanostructured Materials: Science and Technology*, ed. by G.-M. Chow, N.I. Noskova (Kluwer Academic Publishers, Dordrecht Boston London 1998) pp. 207–242
60. I.A. Ovid'ko: Interfaces and misfit defects in nanostructured and polycrystalline films, Rev. Adv. Mater. Sci. **1**, No. 1, 61–107 (2000)
61. I.A. Ovid'ko: Interfaces and stresses in nanostructured and high-T_c superconducting materials. In: *Science and Technology of Interfaces, International Symposium in Honor of Dr. Bhakta Rath*, ed. by S. Ankem, C.S. Pande, I. Ovid'ko, S. Ranganathan (TMS, Warrendale 2002) pp. 245–256
62. I.A. Ovid'ko: Interfacial defects in nanostructures. In: *Encyclopedia of Nanoscience and Nanotechnology*, Chap. 4, ed. by H.S. Nalwa (American Sci. Publ., California 2003)
63. R. Bonnet: On elastic properties of dihedral edge dislocations, Phys. Stat. Sol. (b) **108**, No. 2, 435–441 (1981)
64. S.N.G. Chu: Screw dislocation in a two-phase isotropic thin film, J. Appl. Phys. **53**, No. 4, 3019–3023 (1982)
65. A.L. Kolesnikova: Disclinations and dislocations in a plate of finite thickness. In: *Experimental Investigation and Theoretical Description of Disclinations*, ed. by V.I. Vladimirov (Ioffe Physico-Technical Institute, Leningrad 1984) pp. 194–200, in Russian
66. A.L. Kolesnikova, N.D. Priemskii, A.E. Romanov: *Wedge Straight Disclinations in Elastic Isotropic Plate* (Preprint No. 869 of the Ioffe Physico-Technical Institute, Leningrad 1984) in Russian
67. V.I. Vladimirov, A.L. Kolesnikova, A.E. Romanov: Wedge disclinations in elastic plate, Fizika Metallov i Metallovedenie **60**, No. 6, 1106–1115 (1985) in Russian [English translation in: Phys. Metals Metallogr. **60**, No. 6, 58 (1985)]

68. A.L. Kolesnikova, A.E. Romanov: *Circular Dislocation–Disclination Loops and their Application for Solution of Boundary-Value Problems in Theory of Defects* (Preprint No. 1019 of the Ioffe Physico-Technical Institute, Leningrad 1986) in Russian
69. A.Yu. Belov: Fields of displacements and stresses of rectilinear dislocations in an anisotropic plate, Kristallografiya **32**, No. 3, 550–558 (1987) in Russian [English translation in: Soviet Phys. Crystallogr. **32**, No. 3, 320–324 (1987)]
70. A.L. Kolesnikova, A.E. Romanov: Edge dislocation perpendicular to the surface of a plate, Pis'ma v Jurnal Tekhnicheskoj Fiziki **13**, No. 11, 656–660 (1987) in Russian [English translation in: Soviet Tech. Phys. Lett. **13**, No. 6, 272–274 (1987)]
71. C.L. Malyshev, M.Yu. Gutkin, A.E. Romanov, A.A. Sitnikova, L.M. Sorokin: *Dislocation Models and Diffraction Contrast of Rod-Like Defects in Silicon* (Preprint No. 1109 of the Ioffe Physico-Technical Institute, Leningrad 1987) in Russian
72. S.V. Kamat, J.P. Hirth, B. Carnahan: Image forces on screw dislocations in multilayer structures, Scripta Metall. **21**, No. 11, 1587–1592 (1987)
73. M.L. Öveçoğlu, M.F. Doerner, W.D. Nix: Elastic interactions of screw dislocations in thin films on substrates, Acta Metall. **35**, No. 12, 2947–2957 (1987)
74. L. Stagni, R. Lizzio: Interaction of an edge dislocation with a lamellar inhomogeneity, Mech. Mater. **6**, No. 1, 17–25 (1987)
75. L. Stagni, R. Lizzio: Edge dislocation inside a lamellar inhomogeneity, J. Appl. Phys. **64**, No. 3, 1594–1596 (1988)
76. T.C.T. Ting: Line forces and dislocations in anisotropic elastic composite wedges and spaces, Phys. Stat. Sol. (b) **146**, No. 1, 81–90 (1988)
77. V.G. Gryaznov, A.M. Kaprelov, A.E. Romanov: Critical stability of dislocations in microcrystals, Pis'ma v Jurnal Tekhnicheskoj Fiziki **15**, No. 2, 39–44 (1989) in Russian [English translation in: Soviet Tech. Phys. Lett. **15**, No. 2, 55–57 (1989)]
78. V.G. Gryaznov, A.M. Kaprelov, A.E. Romanov: Size effects of dislocation stability in small particles and microcrystallites, Scr. Metall. **23**, No. 8, 1443–1448 (1989)
79. M.Yu. Gutkin, A.E. Romanov: *Edge Dislocations in Thin Inhomogeneous Plates* (Preprint No. 1407 of the Ioffe Physico-Technical Institute, Leningrad 1989) in Russian
80. M.Yu. Gutkin, A.E. Romanov: *Misfit Dislocations in Thin Two-Phase Heteroepitaxial Plate* (Preprint No. 1408 of the Ioffe Physico-Technical Institute, Leningrad 1989) in Russian
81. M.Yu. Gutkin, M. Militzer, A.E. Romanov, V.I. Vladimirov: Equilibrium position of misfit dislocations, Phys. Stat. Sol. (a) **113**, No. 2, 337–344 (1989)
82. H.O.K. Kirchner: Elastically anisotropic angularly inhomogeneous media. II. A new formalism, Phil. Mag. B **60**, No. 3, 423–432 (1989)
83. R.I. Taylor: The force on a screw dislocation due to a series of layers of alternating shear modulus, Semicond. Sci. Technol. **4**, No. 8, 612–616 (1989)
84. L. Stagni, R. Lizzio: Edge dislocation in a lamellar inhomogeneity with a slipping interface, J. Mater. Sci. **25**, No. 3, 1618–1622 (1990)
85. A.Yu. Belov: The analogue of the Carother's effect and stress anomalies in the theory of disclinations, Phil. Mag. Lett. **64**, No. 4, 207–210 (1991)
86. V.G. Gryaznov, I.A. Polonsky, A.E. Romanov, L.I. Trusov: Size effects of dislocation stability in nanocrystals, Phys. Rev. B **44**, No. 1, 42–46 (1991)

87. V.G. Gryaznov, A.M. Kaprelov, I.A. Polonskii, A.E. Romanov: Disclinations in heterogeneous small particles, Phys. Stat. Sol. (b) **167**, No. 1, 29–36 (1991)
88. V.G. Gryaznov, A.M. Kaprelov, A.E. Romanov, I.A. Polonskii: Channels of relaxation of elastic stresses in pentagonal nanoparticles, Phys. Stat. Sol. (b) **167**, No. 2, 441–450 (1991)
89. M.Yu. Gutkin, A.E. Romanov: Straight edge dislocations in a thin two-phase plate. I. Elastic stress fields, Phys. Stat. Sol. (a) **125**, No. 1, 107–125 (1991)
90. I.A. Polonsky, A.E. Romanov, V.G. Gryaznov, A.M. Kaprelov: Disclination in an elastic sphere, Phil. Mag. A **64**, No. 2, 281–287 (1991)
91. L.I. Trusov, M.Yu. Tanakov, V.G. Gryaznov, A.M. Kaprelov, A.E. Romanov: Relaxation of elastic stresses in overlayed microcrystals, J. Cryst. Growth **114**, Nos. 1–2, 133–140 (1991)
92. A.Yu. Belov: A wedge disclination along the vertex of the wedge-like inhomogeneity in an elastically anisotropic solid, Phil. Mag. A **65**, No. 6, 1429–1444 (1992)
93. M.Yu. Gutkin, A.E. Romanov: Misfit dislocations in a thin two-phase heteroepitaxial plate, Phys. Stat. Sol. (a) **129**, No. 2, 117–126 (1992)
94. M.Yu. Gutkin, A.E. Romanov: Straight edge dislocations in a thin two-phase plate. II. Impurity–vacancy polarization of plate, interaction of a dislocation with interface and free surface, Phys. Stat. Sol. (a) **129**, No. 2, 363–377 (1992)
95. M. Hecker, A.E. Romanov: The stress fields of an edge dislocation near a wedge-shaped boundary, Phys. Stat. Sol. (a) **130**, No. 1, 91–101 (1992)
96. M. Hecker, A.E. Romanov: The stress fields of edge dislocations near wedge-shaped boundaries and bonded wedges, Mater. Sci. Eng. A **164**, Nos. 1–2, 411–414 (1993)
97. A.E. Romanov, I.A. Polonsky, V.G. Gryaznov, S.A. Nepijko, T. Junghaus, N.I. Vitrykhovski: Voids and channels in pentagonal crystals, J. Cryst. Growth **129**, Nos. 3–4, 691–698 (1993)
98. T.J. Gosling, J.R. Willis: The energy of arrays of dislocations in an anisotropic half-space, Phil. Mag. A **69**, No. 1, 65–90 (1994)
99. M.Yu. Gutkin, A.E. Romanov: On the stand-off positions of misfit dislocations, Phys. Stat. Sol. (a) **144**, No. 1, 39–57 (1994)
100. W.J. Yen, C. Hwu: Interaction between dislocations and anisotropic elastic elliptical inclusion, J. Appl. Mech. **61**, No. 3, 548–554 (1994)
101. W.J. Yen, C. Hwu, Y.K. Liang: Dislocation inside, outside, or on the interface of an anisotropic elliptical inclusion, J. Appl. Mech. **62**, No. 2, 306–311 (1995)
102. T.-Y. Zhang: Effect of elastic constants on the critical thickness of an epilayer, Phys. Stat. Sol. (a) **152**, No. 2, 415–429 (1995)
103. T.-Y. Zhang: Anisotropic elasticity study of the critical thickness of an epilayer on a substrate with different elastic constants, J. Appl. Phys. **78**, No. 8, 4948–4957 (1995)
104. R. Bonnet: Elasticity theory of straight dislocations in a multilayer, Phys. Rev. B **53**, No. 16, 10978–10982 (1996)
105. M.Yu. Gutkin, A.E. Romanov, E.C. Aifantis: Screw dislocation near a triple junction of phases with different elastic moduli. I. General solution, Phys. Stat. Sol. (a) **153**, No. 1, 65–75 (1996)
106. M.Yu. Gutkin, A.E. Romanov, E.C. Aifantis: Elastic behaviour of screw dislocations near triple junctions of interphase boundaries, Mater. Sci. Forum **207–209**, Part 2, 605–608 (1996)

107. T.-Y. Zhang: Isotropic and anisotropic elasticity studies of mismatch dislocations, critical thickness and work hardening in epilayers, Defect Diffusion Forum **136–137**, 61–80 (1996)
108. T.-Y. Zhang: Anisotropic elasticity studies of work hardening in epilayers, Scr. Mater. **39**, Nos. 4/5, 473–479 (1998)
109. Sh.-D. Wang: Screw dislocations in a two-phase plate and a sandwich specimen, Phys. Stat. Sol. (b) **215**, No. 2, 933–948 (1999)
110. T.-Y. Zhang, Y.-J. Su: The critical thickness of an epilayer deposited on a semiconductor-on-insulator compliant substrate, Appl. Phys. Lett. **74**, No. 12, 1689–1691 (1999)
111. T.-Y. Zhang, Y.-J. Su: Elasticity studies of the critical thickness of an epilayer deposited on a compliant substrate, Acta. Mater. **47**, No. 4, 1289–1296 (1999)
112. T.-Y. Zhang, S. Lee, L.J. Guido, Ch.-H. Hsueh: Criteria for formation of interface dislocations in a finite thickness epilayer deposited on a substrate, J. Appl. Phys. **85**, No. 11, 7579–7586 (1999)
113. I.A. Ovid'ko: Misfit dislocation walls in solid films, J. Phys.: Cond. Matter **11**, No. 34, 6521–6527 (1999)
114. I.A. Ovid'ko: Misfit disclinations and dislocations at crystal–glass interfaces, Phil. Mag. Lett. **79**, No. 9, 709–713 (1999)
115. I.A. Ovid'ko: Misfit disclinations at crystal/crystal and crystal/glass interfaces, Fizika Tverdogo Tela **41**, No. 9, 1637–1643 (1999) in Russian [English translation in: Phys. Solid State **41**, No. 9, 1500–1505 (1999)]
116. M.Yu. Gutkin, I.A. Ovid'ko: Misfit strains and phase transformations in layered composite solids, J. Phys: Cond. Matter **11**, No. 43, 8607–8616 (1999)
117. I.A. Ovid'ko: Quasiperiodic interfaces and tilt boundaries of finite extent, Mater. Sci. Eng. A **280**, No. 2, 355–358 (2000)
118. D.Yu. Baskov, M.Yu. Gutkin, I.A. Ovid'ko, A.G. Sheinerman: Misfit dislocation walls in multilayered films, Mater. Phys. Mech. **2**, No. 1, 10–14 (2000)
119. M.Yu. Gutkin, I.A. Ovid'ko, A.G. Sheinerman: Misfit dislocations in wire composite solids, J. Phys.: Condens. Matter **12**, No. 25, 5391–5401 (2000)
120. Sh.-D. Wang: Energy of an array of dislocations in a strained epitaxial layer deposited on a finite substrate, J. Appl. Phys. **88**, No. 12, 7089–7094 (2000)
121. Z.M. Xiao, B.J. Chen: A screw dislocation interacting with a coated fiber, Mech. Mater. **32**, No. 8, 485–494 (2000)
122. I.A. Ovid'ko, A.G. Sheinerman: Misfit dislocations in multilayered films on disclinated substrates, J. Phys.: Condens. Matter. **13**, No. 35, 7937–7951 (2001)
123. I.A. Ovid'ko, A.B. Reizis: Grain boundary dislocation climb and diffusion in nanocrystalline solids, Fizika Tverdogo Tela **43**, No. 1, 35–38 (2001) in Russian [English translation in: Phys. Solid State **43**, No. 1, 35–38 (2001)]
124. M.Yu. Gutkin, K.N. Mikaelyan, I.A. Ovid'ko: Generation and evolution of partial misfit dislocations and stacking faults in thin-film heterostructures, Fizika Tverdogo Tela **43**, No. 1, 42–46 (2001) in Russian [English translation in: Phys. Solid State **43**, No. 1, 42–46 (2001)]
125. I.A. Ovid'ko, A.G. Sheinerman: Dislocation dipoles in nanocrystalline films, J. Nanosci. Nanotechnol. **1**, No. 2, 215–220 (2001)
126. A.G. Sheinerman, M.Yu. Gutkin: Misfit disclinations and dislocation walls in a two-phase cylindrical composite, Phys. Stat. Sol. (a) **184**, No. 2, 485–505 (2001)

127. A.G. Sheinerman, M.Yu. Gutkin: Misfit dislocations in a hollow cylindrical film grown on a hole surface, Scr. Mater. **45**, No. 1, 81–87 (2001)
128. A.L. Kolesnikova, A.E. Romanov, I.A. Ovid'ko: Misfit disclination structures in nanocrystalline films, Solid State Phenomena **87**, 265–276 (2002)
129. A.A. Fedorov, A.L. Kolesnikova, I.A. Ovid'ko: Grain boundary dislocation structures near triple junctions in nano- and polycrystalline materials, Mater. Phys. Mech. **5**, No. 1, 49–55 (2002)
130. I.A. Ovid'ko, A.G. Sheinerman: Dislocation dipoles in nanoscale films with compositional inhomogeneities, Philos. Mag. A **82**, No. 16, 3119–3127 (2002)
131. I.A. Ovid'ko, A.G. Sheinerman: Nano-islands on a composite substrate with misfit dislocations, Appl. Phys. A **74**, No. 2, 273–277 (2002)
132. I.A. Ovid'ko: Relaxation mechanisms in strained nanoislands, Phys. Rev. Lett. **88**, No. 4, 046103 (4 pages) (2002)
133. I.A. Ovid'ko, A.G. Sheinerman: Perfect, partial and split dislocations in quantum dots, Phys. Rev. B **66**, 245309 (8 pages) (2002)
134. Zh. Li, J. Shi: The interaction of a screw dislocation with inclusion analyzed by Eshelby equivalent inclusion method, Scr. Mater. **47**, No. 6, 371–375 (2002)
135. N.F. Morozov, I.A. Ovid'ko, Yu.V. Petrov, A.G. Sheinerman: Formation and convergence of nanocracks in mechanically loaded nanocrystalline solids, Rev. Adv. Mater. Sci. **4**, No. 1, 65–69 (2003)
136. M.Yu. Gutkin, I.A. Ovid'ko, A.G. Sheinerman: Misfit dislocations in composites with nanowires, J. Phys.: Condens. Matter **15**, No. 21, 3539–3554 (2003)
137. I.A. Ovid'ko, A.G. Sheinerman: Hyperdislocations in misfit dislocation networks in solid films, J. Phys.: Condens. Matter **15**, No. 12, 2127–2135 (2003)
138. I.A. Ovid'ko, A.G. Sheinerman, N.V. Skiba: Competing relaxation mechanisms in strained semiconducting and superconducting films, J. Phys.: Condens. Matter **15**, No. 8, 1173–1181 (2003)
139. I.A. Ovid'ko, A.G. Sheinerman: Grain boundary dislocations and enhanced diffusion in nanocrystalline bulk materials and films, Philos. Mag. **83**, No. 13, 1551–1563 (2003)
140. A.L. Kolesnikova, A.E. Romanov: Loops of dislocations and disclinations in the method of virtual defects, Fizika Tverdogo Tela **45**, No. 9, 1626–1636 (2003) in Russian [English translation in: Phys. Solid State **45**, No. 9, 1626–1636 (2003)]
141. A.C. Eringen: Screw dislocation in non-local elasticity, J. Phys. D: Appl. Phys. **10**, No. 5, 671–678 (1977)
142. A.C. Eringen: On differential equations of nonlocal elasticity and solutions of screw dislocation and surface waves, J. Appl. Phys. **54**, No. 9, 4703–4710 (1983)
143. A.C. Eringen: Nonlocal continuum theory for dislocations and fracture. In: *The Mechanics of Dislocations*, ed. by E.C. Aifantis, J.P. Hirth (American Society of Metals, Metals Park, Ohio 1985) pp. 101–110
144. F. Gao: Screw dislocation in bi-medium in non-local elasticity, J. Phys. D **23**, No. 3, 328–333 (1990)
145. K.L. Pan: Screw dislocation in semi-infinite non-local hexagonal medium, Radiat. Eff. Defects Solids **133**, No. 2, 167–178 (1995)
146. Y.Z. Povstenko: Circular dislocation loops in non-local elasticity J. Phys. D: Appl. Phys. **28**, No. 1, 105–111 (1995)
147. Y.Z. Povstenko: Straight disclinations in nonlocal elasticity, Int. J. Engng. Sci. **33**, No. 4, 575–582 (1995)

148. Y.Z. Povstenko: Imperfections in non-local elasticity, J. Physique IV **8**, 8, 309–316 (1998)
149. M.Yu. Gutkin, E.C. Aifantis: Screw dislocation in gradient elasticity, Scr. Mater. **35**, No. 11, 1353–1358 (1996)
150. M.Yu. Gutkin, E.C. Aifantis: Edge dislocation in gradient elasticity, Scr. Mater. **36**, No. 1, 129–135 (1997)
151. M.Yu. Gutkin, E.C. Aifantis: Dislocations in the theory of gradient elasticity, Scr. Mater. **40**, No. 5, 559–566 (1999)
152. M.Yu. Gutkin, E.C. Aifantis: Dislocations and disclinations in gradient elasticity, Phys. Stat. Sol. (b) **214**, No. 2, 245–284 (1999)
153. M.Yu. Gutkin, E.C. Aifantis: Dislocations and disclinations in the gradient theory of elasticity, Fizika Tverdogo Tela **41**, No. 12, 2158–2166 (1999) in Russian [English translation in: Phys. Solid State **41**, No. 12, 1980–1988 (1999)]
154. M.Yu. Gutkin, E.C. Aifantis: Nanoscale elastic properties of dislocations and disclinations. In: *Nanostructured Films and Coatings, NATO ARW Series, High Technology*, Vol. 78, ed. by G.-M. Chow, I.A. Ovid'ko, T. Tsakalakos (Kluwer, Dordrecht 2000) pp. 247–254
155. M.Yu. Gutkin, K.N. Mikaelyan, E.C. Aifantis: Behavior of screw dislocations near phase boundaries in the gradient theory of elasticity, Fizika Tverdogo Tela **42**, No. 9, 1606–1612 (2000) in Russian [English translation in: Phys. Solid State **42**, No. 9, 1652–1658 (2000)]
156. K.N. Mikaelyan, M.Yu. Gutkin, E.C. Aifantis: Edge dislocations near phase boundaries in the gradient theory of elasticity, Fizika Tverdogo Tela **42**, No. 9, 1613–1620 (2000) in Russian [English translation in: Phys. Solid State **42**, No. 9, 1659–1667 (2000)]
157. M.Yu. Gutkin, K.N. Mikaelyan, E.C. Aifantis: Screw dislocation near interface in gradient elasticity, Scr. Mater. **43**, No. 6, 477–484 (2000)
158. M.Yu. Gutkin: Nanoscopics of dislocations and disclinations in gradient elasticity, Rev. Adv. Mater. Sci. **1**, No. 1, 27–60 (2000)
159. M.C. Valsakumar, D. Sahoo: Gauge theory of defects in the elastic continuum, Bull. Mater. Sci. **10**, Nos. 1–2, 3–44 (1988)
160. D.G.B. Edelen: A correct, globally defined solution of the screw dislocation problem in the gauge theory of defects, Int. J. Engng. Sci. **34**, No. 1, 81–86 (1996)
161. C. Malyshev: T(3)-gauge model, the Einstein-like gauge equation, and Volterra dislocations with modified asymptotics, Ann. Phys. (N.Y.) **286**, 249–277 (2000)
162. M. Lazar: Dislocation theory as a 3-dimensional gauge theory, Ann. Phys. (Leipzig) **9**, No. 6, 461–473 (2000)
163. M. Lazar: An elastoplastic theory of dislocations as a physical field theory with torsion, J. Phys. A: Math. Gen. **35**, No. 8, 1983–2004 (2002)
164. M. Lazar: Screw dislocations in the field theory of elastoplasticity, Ann. Phys. (Leipzig) **11**, No. 9, 635–649 (2002)
165. M. Lazar: A nonsingular solution of the edge dislocation in the gauge theory of dislocations, J. Phys. A: Math. Gen. **35**, No. 5, 1415–1437 (2002)
166. M. Lazar: Wedge disclination in the field theory of elastoplasticity, Phys. Lett. A **311**, 416–425 (2003)
167. M. Lazar: Twist disclination in the field theory of elastoplasticity, J. Phys.: Cond. Matter **15**, 6781–6800 (2003)

168. U. Herr, J. Jing, R. Birringer, U. Conser, H. Gleiter: Investigation of nanocrystalline iron materials by Mössbauer spectroscopy, Appl. Phys. Lett. **50**, No. 8, 472–474 (1987)
169. H.-E. Schaefer, R. Würschum: Positron lifetime spectroscopy in nanocrystalline iron, Phys. Lett. A **119**, No. 7, 370–374 (1987)
170. R. Würschum, M. Scheytt, H.-E. Schaefer: Nanocrystalline metals and semiconductors studied by positron lifetime spectroscopy, Phys. Stat. Sol. (a) **102**, No. 1, 119–126 (1987)
171. X. Zhu, R. Birringer, U. Herr, H. Gleiter: X-ray diffraction studies of the structure of nanometer-sized crystalline materials, Phys. Rev. B **35**, No. 17, 9085–9090 (1987)
172. H.-E. Schaefer, R. Würschum, R. Birringer, H. Gleiter: Structure of nanometer-sized polycrystalline iron investigated by positron lifetime spectroscopy, Phys. Rev. B **38**, No. 14, 9545–9554 (1988)
173. R.W. Siegel, H. Hahn, S. Ramasamy, L. Zongquan, L. Ting, R. Gronsky: Structures and properties of nanophase TiO_2, J. Phys. Colloq. **49**, No. C-5, 681–686 (1988)
174. R.W. Siegel, S. Ramasamy, H. Hahn, L. Zongquan, L. Ting, R. Gronsky: Synthesis, characterization, and properties of nanophase TiO_2, J. Mater. Res. **3**, No. 6, 1367–1372 (1988)
175. T. Haubold, R. Birringer, B. Lengeler, H. Gleiter: EXAFS studies of nanocrystalline materials exhibiting a new solid state structure, Phys. Lett. A **135**, Nos. 8–9, 461–466 (1989)
176. K. Hayashi, H. Etoh: Pressure sintering of iron, cobalt, nickel and copper ultrafine powders and the crystal grain size and hardness of the compacts, Mater. Trans. Jpn. Inst. Metals **30**, No. 11, 925–931 (1989)
177. E. Jorra, H. Franz, J. Peisl, G. Wallner, W. Petry, T. Haubold, R. Birringer, H. Gleiter: Small-angle neutron scattering from nanocrystalline Pd, Phil. Mag. B **60**, No. 2, 159–168 (1989)
178. G. McMahon, U. Erb: Structural transitions in electroplated Ni–P alloy, J. Mater. Sci. Lett. **8**, No. 7, 865–868 (1989)
179. J.S.C. Jang, C.C. Koch: Amorphization and disordering of the Ni_3Al ordered intermetallic by mechanical milling, J. Mater. Res. **5**, No. 3, 498–510 (1990)
180. G. Palumbo, U. Erb, K.T. Aust: Triple line disclination effects on the mechanical behaviour of materials, Scr. Metall. Mater. **24**, No. 12, 2347–2350 (1990)
181. W. Wunderlich, Y. Ishida, R. Maurer: HREM studies of the microstructure of nanocrystalline palladium, Scr. Metall. Mater. **24**, No. 2, 403–408 (1990)
182. H. Chang, H.J. Höfler, C.J. Altstetter, R.S. Averback: Synthesis, processing and properties of nanophase TiAl, Scr. Metall. Mater. **25**, No. 5, 1161–1166 (1991)
183. T. Christman, M. Jain: Processing and consolidation of bulk nanocrystalline titanium-aluminide, Scr. Metall. Mater. **25**, No. 4, 767–772 (1991)
184. D.G. Morris, M.A. Morris: Microstructure and strength of nanocrystalline copper alloy prepared by mechanical alloying, Acta Metall. Mater. **39**, No. 8, 1763–1770 (1991)
185. R.Z. Valiev, N.A. Krasilnikov, N.K. Tsenev: Plastic deformation of alloys with submicron-grained structure, Mater. Sci. Engng. A **137**, 35–40 (1991)
186. V.G. Gryaznov, M.Yu. Tanakov, L.I. Trusov: Plasticity and mass-transfer in contacting nanoparticles, J. Mater. Sci. **27**, No. 18, 4829–4841 (1992)

187. R.Z. Valiev, F. Chmelik, F. Bordeaux, G. Kapelski, B. Baudelet: The Hall-Petch relation in submicro-grained Al–1.5%Mg alloy, Scr. Metall. Mater. **27**, No. 7, 855–860 (1992)
188. G.E. Fougere, J.R. Weertman, R.W. Siegel: On the hardening and softening of nanocrystalline materials, Nanostruct. Mater. **3**, Nos. 1–6, 379–384 (1993)
189. M. Ke, W.W. Milligan, S.A. Hackney, J.E. Carsley, E.C. Aifantis: HREM study of fracture and deformation behavior of nanostructured thin films, Mat. Res. Soc. Symp. Proc. **308**, 565–569 (1993)
190. N.P. Kobelev, Ya.M. Soifer, R.A. Andrievski, B. Gunther: Microhardness and elastic properties of nanocrystalline silver, Nanostruct. Mater. **2**, No. 5, 537–544 (1993)
191. J. Languillaume, F. Chmelik, G. Kapelski, F. Bordeaux, A.A. Nazarov, G. Canova, C. Esling, R.Z. Valiev, B. Baudelet: Microstructures and hardness of ultrafine-grained Ni_3Al, Acta Metall. Mater. **41**, No. 10, 2953–2962 (1993)
192. X.D. Liu, B.Z. Ding, Z.Q. Hu, K. Lu, Y.Z. Wang: Properties of nanocrystalline Fe–Cu–Si–B alloys generated by crystallization of the amorphous alloy, Physica B **192**, No. 4, 345–350 (1993)
193. X.D. Liu, Z.Q. Hu, B.Z. Ding: Hall–Petch relation in nanocrystalline Fe–Mo Si–B alloys, Nanostruct. Mater. **2**, No. 5, 545–552 (1993)
194. G.E. Fougere, J.R. Weertman, R.W. Siegel: Processing and mechanical behavior of nanocrystalline Fe, Nanostruct. Mater. **5**, No. 2, 127–134 (1995)
195. M. José-Yacamán: High resolution electron microscopy of nanostructured materials, Nanostruct. Mater. **5**, No. 2, 171–178 (1995)
196. M. Ke, S.A. Hackney, W.W. Milligan, E.C. Aifantis: Observation and measurement of grain rotation and plastic strain in nanostructured metal thin films, Nanostruct. Mater. **5**, No. 6, 689–697 (1995)
197. A.A. Nazarov, A.E. Romanov, R.Z. Valiev: Models of the defect structure and analysis of the mechanical behavior of nanocrystals, Nanostruct. Mater. **6**, Nos. 5–8, 775–778 (1995)
198. D.H. Ping, T.S. Xie, D.X. Li, H.Q. Ye: High resolution transmission electron microscopy studies of the microstructure of nanocrystalline $Ti_{70}Ni_{20}Si_{10}$ alloy, Nanostruct. Mater. **5**, No. 4, 457–464 (1995)
199. X.Y. Qin, X.J. Wu, L.D. Zhang: The microhardness of nanocrystalline silver, Nanostruct. Mater. **5**, No. 1, 101–110 (1995)
200. T.D. Shen, C.C. Koch: Formation and hardening effects in nanocrystalline Ti–N alloys prepared by mechanical alloying, Nanostruct. Mater. **5**, No. 6, 615–629 (1995)
201. T.R. Smith, K.S. Vecchio: Synthesis and mechanical properties of nanoscale mechanically-milled NiAl, Nanostruct. Mater. **5**, No. 1, 11–23 (1995)
202. A.H. King: The geometric and thermodynamic properties of grain boundary junctions, Interf. Sci. **7**, Nos. 3–4, 251–271 (1999)
203. V.B. Rabukhin: On the mechanism of action of boundary triple junction in the processes of grain boundary inelasticity, Poverkhnost'. Fizika, Khimiya, Mekhanika **7**, 126–130 (1986) in Russian
204. G. Palumbo, K.T. Aust: Triple-line corrosion in high-purity nickel, Mater. Sci. Eng. A **113**, 139–147 (1989)
205. A.V. Osipov, I.A. Ovid'ko: Diffusion-induced decay of disclinations and solid state amorphization in mechanically alloyed materials, Appl. Phys. A **54**, No. 3, 517–519 (1992)

206. M.Yu. Gutkin, I.A. Ovid'ko: Disclinations, amorphization and microcrack generation at grain boundary junctions in polycrystalline solids, Phil. Mag. A **70**, No. 4, 561–575 (1994)
207. K.M. Yin, A.H. King, T.E. Hsieh, F.R. Chen, J.J. Kai, L. Chang: Segregation of bismuth to triple junctions in copper, Microscopy and Microanalysis **3**, 417–422 (1997)
208. U. Czubayko, V.G. Sursaeva, G. Gottstein, L.S. Shvindlerman: Steady state motion of triple junctions: Theory and experiment. In: *Grain Growth in Polycrystalline Materials III*, ed. by H. Weiland, B.L. Adams, A.D. Rollett (TMS, Pittsburg 1998) pp. 423–430
209. G. Gottstein, A.H. King, L.S. Shvindlerman: The effect of triple junction drag on grain growth, Acta Mater. **48**, No. 2, 397–403 (2000)
210. K. Owusu-Boahen, A.H. King: The early stages of plastic yielding in polycrystalline gold thin films, Acta Mater. **49**, No. 2, 237–247 (2001)
211. A.A. Fedorov, M.Yu. Gutkin, I.A. Ovid'ko: Triple junction diffusion and plastic flow in fine-grained materials, Scr. Mater. **47**, No. 1, 51–55 (2002)
212. J. Horvath, R. Birringer, H. Gleiter: Diffusion in nanocrystalline materials, Solid State Comm. **62**, 319 (1987)
213. H.-E. Schaefer, R. Würschum, T. Gessmann, G. Stöckl, P. Scharwaechter, W. Frank, R.Z. Valiev, H.-J. Fecht, C. Moelle: Diffusion and free volumes in nanocrystalline Pd, Nanostruct. Mater. **6**, Nos. 5–8, 869–872 (1995)
214. Yu.R. Kolobov, G.P. Grabovetskaya, I.V. Ratochka, K.V. Ivanov: Diffusion-induced creep of polycrystalline and nanostructured metals, Nanostruct. Mater. **12**, Nos. 5–8, 1127–1130 (1999)
215. Yu.R. Kolobov, G.P. Grabovetskaya, K.V. Ivanov, R.Z. Valiev, T.C. Lowe: Copper grain boundary diffusion and diffusion induced creep in nanostructured nickel. In: *Investigations and Applications of Severe Plastic Deformation*, NATO Science Ser., ed. by T.C. Lowe, R.Z. Valiev (Kluwer, Dordrecht 2000) p. 261
216. J.R. Weertman, D. Farkas, K. Hemker, H. Kung, M. Mayo, R. Mitra, H. Van Svygenhoven: Structure and mechanical behavior of bulk nanocrystalline materials, MRS Bull. **24**, No. 2, 44–50 (1999)
217. P. Rodriguez, D. Sundararaman, R. Divakar, V.S. Raghunathan: Structure of grain boundaries in nanocrystalline and quasicrystalline materials, Chemistry for Sustainable Development **8**, 69–72 (2000)
218. M. Murayama, J.M. Howe, H. Hidaka, S. Takaki: Atomic-level observation of disclination dipoles in mechanically milled, nanocrystalline Fe, Science **295**, No. 5564, 2433–2435 (2002)
219. M. Chen, E. Ma, K.J. Hemker, H. Sheng, Y. Wang, X. Cheng: Deformation twinning in nanocrystalline aluminum, Science **300**, No. 5623, 1275–1277 (2003)
220. R. Hull, A.T. Fiory, J.C. Bean, J.M. Gibson, L. Scott, J.L. Benton, S. Nakahara: The structure of Ge_xSi_{1-x}/Si(100) interfaces and superlattices. In: *Proc. 13th Intern. Conf. on Defects in Semiconductors, Coronado, Calif., USA, August 12–17, 1984* (Metall. Soc. AIME, Warrendale 1985) pp. 505–511
221. N. Hayafuji, S. Ochi, M. Miyashita, M. Tsugami, T. Muratani, A. Kawagishi: Effectiveness of AlGaAs/GaAs superlattices in reducing dislocation density in GaAs on Si, J. Cryst. Growth **93**, Nos. 1–4, 494–498 (1988)

222. N.A. El-Masry, J.C. Tarn, N.H. Karam: Interactions of dislocations in GaAs grown on Si substrates with InGaAs–GaAsP strained layered superlattices, J. Appl. Phys. **64**, No. 7, 3672–3677 (1988)
223. A.K. Gutakovskii, Yu.O. Kanter, V.Yu. Karasev, N.A. Kiselev, S.M. Pintus, S.V. Rubanov, S.I. Stenin, A.A. Fedorov: Structure of films and interfaces in multilayer systems based on indium and gallium arsenides, Doklady AN SSSR **304**, No. 1–3, 355–357 (1989) in Russian [English translation in: Sov. Phys. Dokl. **34**, No. 1, 3–4 (1989)]
224. R.W. Siegel, G.E. Fougere: Mechanical properties of nanophase metals, Nanostruct. Mater. **6**, Nos. 1–4, 205–216 (1995)
225. H. Hahn, K.A. Padmanabhan: Mechanical response of nanostructured materials, Nanostruct. Mater. **6**, Nos. 1–4, 191–200 (1995)
226. H.W. Song, S.R. Guo, Z.Q. Hu: A coherent polycrystal model for the inverse Hall–Petch relation in nanocrystalline materials, Nanostruct. Mater. **11**, No. 2, 203–210 (1999)
227. M.J. Mayo: High and low temperature superplasticity in nanocrystalline materials, Nanostruct. Mater. **9**, Nos. 1–8, 717–726 (1997)
228. M.J. Mayo: Nanocrystalline ceramics for structural applications: Processing and properties. In: *Nanostructured Materials: Science and Technology*, ed. by G.-M.Chow, N.I. Noskova (Dordrecht, Kluwer 1998) pp. 361–385
229. E.O. Hall: Deformation and ageing of mild steel, Proc. Phys. Soc. London B **64**, No. 1, 747–753 (1951)
230. N.J. Petch: The cleavage strength of polycrystals, J. Iron Steel Inst. **174**, 25–28 (1953)
231. R.A. Masumura, P.M. Hazzledine, C.S. Pande: Yield stress of fine grained materials, Acta Mater. **46**, No. 13, 4527–4534 (1998)
232. C.S. Pande, R.A. Masumura, R.W. Armstrong: Pile-up based Hall–Petch relation for nanoscale materials, Nanostruct. Mater. **2**, 323–331 (1993)
233. P.G. Sanders, J.A. Eastman, J.A. Weertman: Elastic and tensile behavior of nanocrystalline copper and palladium, Acta Mater. **45**, No. 10, 4019–4025 (1997)
234. R.M. Christensen: *Mechanics of Composite Materials* (John Wiley, New York 1979)
235. S. Nemat-Nasser, M. Hori: *Micromechanics: Overall Properties of Heterogeneous Materials* (Elsevier Publishers B.V., Amsterdam London New York Tokyo 1993)
236. U.F. Kocks: The relation between polycrystal deformation and single crystal deformation, Metal. Trans. **1**, No. 5, 1121–1143 (1970)
237. A. Lasalmonie, J.L. Strudel: Influence of grain size on the mechanical behaviour of some high strength materials, J. Mater. Sci. **21**, No. 6, 1837–1852 (1986)
238. J.S.C. Jang, C.C. Koch: The Hall–Petch relationship in nanocrystalline iron produced by ball milling, Scr. Metall. Mater. **24**, No. 8, 1599–1604 (1990)
239. V.G. Gryaznov, M.Yu. Gutkin, A.E. Romanov, L.I. Trusov: On the yield stress of nanocrystals, J. Mater. Sci. **28**, No. 16, 4359–4365 (1993)
240. G.D. Hughes, S.D. Smith, C.S. Pande, H.R. Johnson, R.W. Armstrong: Hall–Petch strengthening for the microhardness of twelve nanometer grain diameter electrodeposited nickel, Scr. Metall. **20**, No. 1, 93–97 (1986)

241. H.J. Höfler, R.S. Averback: Grain growth in nanocrystalline TiO_2 and its relation to Vickers hardness and fracture toughness, Scr. Metall. Mater. **24**, No. 12, 2401–2406 (1990)
242. A.H. Chokshi, A. Rosen, J. Karch, H. Gleiter: On the validity of the Hall–Petch relationship in nanocrystalline materials, Scr. Metall. **23**, No. 10, 1679–1684 (1989)
243. G.W. Nieman, J.R. Weertman, R.W. Siegel: Microhardness of nanocrystalline palladium and copper produced by inert-gas condensation, Scr. Metall. **23**, No. 12, 2013–2018 (1989)
244. G.W. Nieman, J.R. Weertman, R.W. Siegel: Mechanical behavior of nanocrystalline Cu and Pd, J. Mater. Res. **6**, No. 5, 1012–1027 (1991)
245. K. Lu, W.D. Wei, J.T. Wang: Microhardness and fracture properties of nanocrystalline Ni–P alloy, Scr. Metall. Mater. **24**, No. 12, 2319–2323 (1990)
246. J. Friedel: *Dislocations* (Pergamon Press, Oxford 1964)
247. H. Gleiter: Nanocrystalline materials: state of art and perspectives, Nanostruct. Mater. **6**, Nos. 1–4, 3–14 (1995)
248. P.G. Cheremskoj, V.V. Slezov, V.I. Betekhtin: *Pores in Solid Bodies* (Energoatomizdat, Moscow 1990) in Russian
249. R.W. Armstrong: The influence of polycrystal grain size on several mechanical properties of materials, Metall. Trans. **1**, No. 5, 1169–1176 (1970)
250. V.I. Trefilov, V.F. Moiseev, E.P. Pechkovsky: *Strain Hardening and Fracture of Polycrystalline Metals* (Naukova Dumka, Kiev 1987) in Russian
251. R.W. Armstrong: Strength of ultrafine-grain metals. In: *Ultrafine-Grain Metals*, ed. by J.J. Burke, V. Weiss (Syracuse University Press, Syracuse, NY 1972)
252. N.I. Noskova, E.G. Ponomareva, I.A. Pereturina, V.N. Kuznetsov: Strength and plasticity of a Pd–Cu–Si alloy in the amorphous and nanocrystalline states, Fizika Metallov i Metallovedenie **81**, No. 1, 162–170 (1996) in Russian [English translation in: Phys. Metals Metallogr. **81**, No. 1, 110 (1996)]
253. N.I. Noskova: Structure and properties of nanocrystalline multiphase alloys. In: *Structure, Phase Transformations and Properties of Nanocrystalline Alloys* (Ural Department of Russian Academy of Sciences, Ekaterinburg 1997) pp. 5–20, in Russian
254. N.I. Noskova: The nanocrystalline alloys: the structure and properties. In: *Nanostructured Materials: Science and Technology*, ed. by G.-M. Chow, N.I. Noskova (Kluwer Academic Publishers, Dordrecht Boston London 1998) pp. 93–119
255. M.Yu. Gutkin, I.A. Ovid'ko: Disclinations and yield stress of metallic glass–nanocrystal composites, Nanostruct. Mater. **2**, No. 6, 631–636 (1993)
256. M.Yu. Gutkin, I.A. Ovid'ko: On the role of disclinations in deformation processes in polycrystals and metallic glass–nanocrystal composites. In: *Strength of Materials, Proc. 10th Int. Conf. (ICSMA-10), Sendai, Japan, Aug. 21–26, 1994*, ed. by H. Oikawa, K. Maruyama, S. Takeuchi, M. Yamaguchi (The Japan Institute of Metals, Sendai 1994) pp. 227–230
257. J.E. Carsley, J. Ning, W.W. Milligan, S.A. Hackney, E.C. Aifantis: A simple, mixtures-based model for the grain size dependence of strength in nanophase metals, Nanostruct. Mater. **5**, No. 4, 441–448 (1995)
258. H.S. Kim: A composite model for mechanical properties of nanocrystalline materials, Scr. Mater. **39**, No. 8, 1057–1061 (1998)

259. D.A. Konstantinidis, E.C. Aifantis: On the 'anomalous' hardness of nanocrystalline materials, Nanostruct. Mater. **10**, No. 7, 1111–1118 (1998)
260. I.A. Ovid'ko: *Defects in Condensed Media: Glasses, Crystals, Quasicrystals, Liquid Crystals, Magnetics, Superfluids* (Znanie, Leningrad, 1991) in Russian
261. V.A. Likhachev, V.E. Shudegov: *Principles of Organization of Amorphous Structures* (St. Petersburg State University, St. Petersburg 1999) in Russian
262. N. Rivier: Continuous random network from graphs to glasses, Adv. Phys. **36**, No. 1, 95–134 (1987)
263. V.I. Vladimirov, A.E. Romanov: *Disclinations in Crystals* (Nauka, Leningrad 1986) in Russian
264. V.V. Rybin: *Large Plastic Deformations and Fracture of Metals* (Metallurgiya, Moscow 1986) in Russian
265. M. Seefeldt: Disclinations in large-strain plastic deformation and work-hardening, Rev. Adv. Mater. Sci. **2**, No. 1, 44–79 (2001)
266. P. Klimanek, A.E. Romanov, M. Seefeldt (Eds.): *Local Lattice Rotations and Disclinations in Microstructures of Distorted Crystalline Materials*, Solid State Phenomena **87**, 1–310 (2002)
267. M.Yu. Gutkin, I.A. Ovid'ko, A.E. Romanov: Intersection of dislocations with disclinations and flow stress in metallic glasses, Radiat. Eff. Def. Sol. **129**, Nos. 3–4, 239–255 (1994)
268. G.S. Pisarenko, A.P. Yakovlev, V.V. Matveev: *Handbook on Strength of Materials* (Naukova Dumka, Kiev 1988) in Russian
269. D. Tabor: *The Hardness of Metals* (Oxford University Press, London 1951) p. 107
270. R.C. Morris: Disclination–dislocation model of metallic glass structures, J. Appl. Phys. **50**, No. 5, 3250–3257 (1979)
271. C.-P. Chou, L.A. Davis, M.C. Narasimhan: Elastic constants of metallic glasses, Scr. Metall. **11**, No. 5, 417–423 (1977)
272. L.A. Davis, Y.T. Yeow, P.M. Anderson: Bulk stiffnesses of metallic glasses, J. Appl. Phys. **53**, No. 7, 4834–4837 (1982)
273. A.M. El-Sherik, U. Erb, G. Palumbo, K.T. Aust: Deviations from Hall–Petch behaviour in as-prepared nanocrystalline nickel, Scr. Metall. Mater. **27**, No. 9, 1185–1188 (1992)
274. R.A. Jago, N. Hansen: Grain-size effects in the deformation of polycrystalline iron, Acta Metall. Mater. **34**, No. 9, 1711–1720 (1986)
275. N. Behnood, R.M. Douthwaite, J.T. Evans: The yield and flow stress of Cu 1%Cd alloy, Acta Metall. **28**, No. 8, 1133–1142 (1980)
276. G.E. Fougere, J.R. Weertman, R.W. Siegel, S. Kim: Grain-size dependent hardening and softening of nanocrystalline Cu and Pd, Scr. Metall. Mater. **26**, No. 12, 1879–1883 (1992)
277. N. Wang, Z. Wang, K.T. Aust, U. Erb: Effect of grain size on the mechanical properties of nanocrystalline materials, Acta Metall. Mater. **43**, 519 (1995)
278. T.G. Nieh, J. Wadsworth: Hall–Petch relation in nanocrystalline solids, Scr. Metall. Mater. **25**, No. 4, 955–958 (1991)
279. G.W. Nieman, J.R. Weertman, R.W. Siegel: Mechanical behavior of nanocrystalline metals, Nanostruct. Mater. **1**, No. 2, 185–190 (1992)
280. J.R. Weertman: Hall–Petch strengthening in nanocrystalline metals, Mater. Sci. Eng. A **166**, Nos. 1–2, 161–167 (1993)

281. C.J. Youngdahl, P.G. Sanders, J.A. Eastman, J.R. Weertman: Compressive yield strengths of nanocrystalline Cu and Pd, Scr. Mater. **37**, No. 6, 809–813 (1997)
282. H.S. Kim, Y. Estrin, M.B. Bush: Plastic deformation behaviour of fine-grained materials, Acta Mater. **48**, No. 2, 493–504 (2000)
283. I.A. Ovid'ko: Quasi-nanocrystalline materials, Nanostruct. Mater. **8**, No. 2, 149–153 (1997)
284. I.A. Ovid'ko: Quasiperiodic interfaces and tilt boundaries of finite extent, Mater. Sci. Eng. A **280**, No. 2, 355–358 (2000)
285. Y. Estrin: Dislocation density-related constitutive modeling. In: *Unified Constitutive Laws of Plastic Deformation*, ed. by A.S. Krausz, K. Krausz (Academic Press, New York 1996) pp. 69–106
286. P.G. Sanders, J.A. Eastman, J.R. Weertman: Tensile behavior of nanocrystalline copper. In: *Processing and Properties of Nanocrystalline Materials*, ed. by C. Suryanarayana, J. Singh, F.H. Froes (TMS, Warrendale, PA 1996) pp. 379–386
287. R. Suryanarayana, C.A. Frey, S.M.L. Sastry, B.E. Waller, S.E. Bates, W.E. Buhro: Mechanical properties of nanocrystalline copper produced by solution-phase synthesis, J. Mater. Res. **11**, No. 2, 439–448 (1996)
288. J.C.M. Li, Y.T. Chou: The role of dislocations in the flow stress grain size relationships, Metall. Trans. **1**, No. 5, 1145–1159 (1970)
289. T.E. Mitchell, S.S. Hecker, R.L. Smialek: Dislocation pile-ups in anisotropic crystals, Phys. Stat. Sol. **11**, No. 2, 585–594 (1965)
290. R.W. Armstrong, A.K. Head: Dislocation quenching and fracture in an elastically anisotropic material, Acta Metall. **13**, No. 7, 759–764 (1965)
291. C.S. Pande, R.A. Masumura: Effect of elastic anisotropy on the brittle-fracture of a polycrystalline solid. In: *Advances in Fracture Research (Proc. of Sixth Intern. Conf. on Fracture, New Delhi, India, Dec. 4–10, 1984)*. Vols. 1–6, ed. by S.P. Valluri, D.M.R. Taplin, P.R. Rao, J.F. Knott, R. Dubey (Pergamon Press, Oxford 1986) pp. 857–862
292. N. Louat: Alloys, strong at room and elevated temperatures from powder metallurgy, Acta Metall. 33, No. 1, 59–69 (1985)
293. A.G. Evans, J.P. Hirth: Deformation of nanoscale cermets, Scr. Metall. Mater. **26**, No. 11, 1675–1680 (1992)
294. C.S. Pande, R.A. Masumura: *Processing and Properties of Nanocrystalline Materials*, ed. by C. Suryanarayana, J. Singh, F.H. Froes (TMS, Warrendale, PA 1996) p. 387
295. Y.T. Chou: Dislocation pile-ups against a locked dislocation of a different Burgers vector, J. Appl. Phys. **38**, No. 5, 2080–2085 (1967)
296. R.Z. Valiev, E.V. Kozlov, Yu.F. Ivanov, J. Lian, A.A. Nazarov, B. Baudelet: Deformation behavior of ultrafine-grained copper, Acta Mater. **42**, 2467–2473 (1994)
297. G.A. Malygin: Breakdown of the Hall–Petch law in micro- and nanocrystalline materials, Fizika tverdogo tela **37**, No. 8, 2281–2292 (1995) in Russian [English translation in: Phys. Solid State **37**, No. 8, 1248–1253 (1995)]
298. J.C.M. Li: Petch relation and grain boundary sources, Trans. TMS-AIME **227**, No. 1, 239–247 (1963)
299. K. Lu, M.L. Sui: An explanation to the abnormal Hall–Petch relation in nanocrystalline materials, Scr. Metall. Mater. **28**, 1465–1470 (1993)

300. R.O. Scattergood, C.C. Koch: A modified model for Hall–Petch behavior in nanocrystalline materials, Scr. Mater. **27**, No. 9, 1195–1200 (1992)
301. A.A. Nazarov, A.E. Romanov, R.Z. Valiev, B. Baudelet: The role of internal stresses in the deformation behaviour of nanocrystals. In: *Strength of Materials: Fundamental Physical Aspects of the Strength of Crystalline Materials (ICSMA-10), Proc. 10th Int. Conf., Sendai, Japan, Aug. 21–26, 1994*, ed. by H. Oikawa, K. Maruyama, S. Takeuchi, M. Yamaguchi (The Japan Institute of Metals, Sendai 1994) pp. 877–879
302. A.A. Nazarov: On the pile-up model of the grain size-yield stress relation for nanocrystals, Scr. Mater. **34**, No. 5, 697–701 (1996)
303. J. Lian, B. Baudelet, A.A. Nazarov: Model for the prediction of the mechanical behaviour of nanocrystalline materials, Mater. Sci. Eng. A **172**, 23–29 (1993)
304. S.G. Zaichenko, A.M. Glezer: Disclination mechanism for plastic deformation of nanocrystalline materials, Fizika tverdogo tela **39**, No. 11, 2023–2028 (1997) in Russian [English translation in: Phys. Solid State **39**, No. 11, 1810–1814 (1997)]
305. A.E. Romanov: Continuum theory of defects in nanoscaled materials, Nanostruct. Mater. **6**, Nos. 1–4, 125–134 (1995)
306. R.W. Siegel: Nanophase materials. In: *Encyclopedia of Applied Physics*, Vol. 11, ed. by G.L. Trigg (VCH, Weinheim 1994) pp. 173–200
307. H. Hahn, P. Mondal, K.A. Padmanabhan: Plastic deformation of nanocrystalline materials, Nanostruct. Mater. **9**, Nos. 1–8, 603–606 (1997)
308. V.V. Astanin, A.V. Sisanbaev, A.I. Pshenichnyuk, O.A. Kaibyshev: Self-organization of cooperative grain boundary sliding in aluminium tricrystals, Scr. Mater. **36**, No. 1, 117–122 (1997)
309. A.A. Fedorov, M.Yu. Gutkin, I.A. Ovid'ko: Transformations of grain boundary dislocation pile-ups in nano- and polycrystalline materials, Acta Mater. **51**, No. 4, 887–898 (2003)
310. H.V. Van Swygenhoven, M. Spaczer, A. Caro, D. Farkas: Competing plastic deformation mechanisms in nanophase metals, Phys. Rev. B **60**, No. 1, 22–25 (1999)
311. H.V. Van Swygenhoven, M. Spaczer, A. Caro: Microscopic description of plasticity in computer generated metallic nanophase samples: A comparison between Cu and Ni, Acta Mater. **47**, No. 10, 3117–3126 (1999)
312. J. Schiotz, T. Vegge, F.D. Di Tiolla, K.W. Jakobsen: Atomic-scale simulations of the mechanical deformation of nanocrystalline metals, Phys. Rev. B **60**, No. 17, 11971–11983 (1999)
313. V. Yamakov, D. Wolf, M. Salazar, S.R. Phillpot, H. Gleiter: Length-scale effects in the nucleation of extended lattice dislocations in nanocrystalline Al by molecular-dynamics simulation, Acta Mater. **49**, No. 14, 2713–2722 (2001)
314. V. Yamakov, D. Wolf, S.R. Phillpot, H. Gleiter: Grain-boundary diffusion creep in nanocrystalline palladium by molecular-dynamics simulation, Acta Mater. **50**, No. 1, 61–73 (2002)
315. V. Yamakov, D. Wolf, S.R. Phillpot, A.K. Mukherjee, H. Gleiter: Dislocation processes in the deformation of nanocrystalline aluminium by molecular-dynamics simulation, Nature Materials **1**, 45–48 (2002)
316. B. Burton: Interface reaction controlled diffusional creep: A consideration of grain-boundary dislocation climb sources, Mat. Sci. Eng. **10**, No. 1, 9–14 (1972)

317. K.J. Kurzydlowski: A model for the flow stress dependence on the distribution of grain size in polycrystals, Scr. Metall. Mater. **24**, No. 5, 879–883 (1990)
318. G. MacHahon, U. Erb: Bulk amorphous and nanocrystalline Ni–P alloys by electroplating. In: *Microstructured Science*, Vol. 17, ed. by P.J. Kenny et al. (ASM, Metals Park, OH 1989) p. 447
319. M.Yu. Gutkin, I.A. Ovid'ko, C.S. Pande: Yield stress of nanocrystalline materials: Role of grain boundary dislocations, triple junctions and Coble creep, Phil. Mag. (2004)
320. F.A. Mohamed, Y. Li: Creep and superplasticity in nanocrystalline materials: Current understanding and future prospects, Mater. Sci. Eng. A **298**, 1–15 (2001)
321. T. Volpp, E. Göring, W.-M. Kuschke, E. Arzt: Grain size determination and limits to Hall–Petch behavior in nanocrystalline NiAl powders, Nanostruct. Mater. **8**, No. 7, 855–865 (1997)
322. H.S. Kim, Y. Estrin, M.B. Bush: Constitutive modeling of strength and plasticity of nanocrystalline metallic materials, Mater. Sci. Eng. A **316**, No. 1, 195–199 (2001)
323. A.B. Witney, P.G. Sanders, J.R. Weertman, J.A. Eastman: Fatigue of nanocrystalline copper, Scr. Metall. Mater. **33**, No. 12, 2025–2030 (1995)
324. R.A. Andrievskii, G.V. Kalinnikov, N.P. Kobelev, Ya.M. Soifer, D.V. Shtansky: Structure and physicochemical properties of nanocrystalline boride–nitride films, Fizika tverdogo tela **39**, No. 10, 1859–1864 (1997) in Russian [English translation in: Phys. Solid State **39**, No. 10, 1661–1666 (1997)]
325. R.A. Andrievskii: The state-of-the-art of nanostructured high melting point compound-based materials. In: *Nanostructured Materials: Science and Technology*, NATO ASI Ser., Vol. 50, ed. by G.M. Chow, N.I. Noskova (Kluwer Academic Publ., Dordrecht 1998) pp. 263–282
326. K.A. Padmanabhan, G.J. Davies: *Superplasticity* (Springer, Berlin 1980)
327. I.I. Novikov, V.K. Portnoy: *Superplasticity in Alloys with Ultrafine Grain Structure* (Metallurgiya, Moscow 1981) in Russian
328. O.A. Kaibyshev, F.Z. Utyashev: *Superplasticity, Structure Refinement and Treatment of Hard-Deforming Alloys* (Nauka, Moscow 2002) in Russian
329. R.W. Siegel: Is superplasticity in the future of nanophase materials? Mater. Res. Soc. Symp. Proc. **196**, 59–70 (1990)
330. J. Pilling, N. Ridley: *Superplasticity in Crystalline Solids* (The Institute of Metals, London 1989)
331. A.K. Mukherjee: Superplasticity in metals, ceramics and intermetallics. In: *Materials Science and Technology*, Vol. 6, ed. by R.W. Cahn, P. Haasen, E.J. Kramer (VCH Publishers, Weinheim 1993) p. 407
332. M.G. Zelin, A.K. Mukherjee: Geometrical aspects of superplastic flow, Mater. Sci. Eng. A **208**, No. 2, 210–225 (1996)
333. G.A. Salishchev, O.R. Valiakhmetov, R.M. Galeyev: Formation of submicrocrystalline structure in the titanium alloy VT8 and its influence on mechanical properties, J. Mater. Sci. **28**, No. 11, 2898–2902 (1993)
334. M. Jain, T. Christman: Synthesis, processing, and deformation of bulk nanophase $Fe_{28}Al_2Cr$ intermetallic, Acta Metall. Mater. **42**, No. 6, 1901–1911 (1994)
335. R.S. Mishra, R.Z. Valiev, A.K. Mukherjee: The observation of tensile superplasticity in nanocrystalline materials, Nanostruct. Mater. **9**, Nos. 1–8, 473–476 (1997)

336. S.X. McFadden, R.S. Mishra, R.Z. Valiev, A.P. Zhylyaev, A.K. Mukherjee: Low-temperature superplasticity in nanostructured nickel and metal alloys, Nature **398**, No. 6729, 684–686 (1999)
337. R.S. Mishra, R.Z. Valiev, S.X. McFadden, R.K. Islamgaliev, A.K. Mukherjee: High-strain-rate superplasticity from nanocrystalline Al alloy 1420 at low temperatures, Phil. Mag. A **81**, No. 1, 37–48 (2001)
338. A.K. Mukherjee: An examination of the constitutive equation for elevated temperature plasticity, Mater. Sci. Eng. A **322**, 1–22 (2002)
339. V.V. Astanin, O.A. Kaibyshev, S.N. Faizova: Cooperative grain-boundary sliding under superplastic flow, Scr. Metall. Mater. **25**, No. 12, 2663–2668 (1991)
340. H.S. Yang, M.G. Zelin: Strain induced morphological changes of α_2- and β-phases in Ti_3Al alloys during superplastic deformation, Scr. Metall. Mater. **26**, No. 11, 1707–1712 (1992)
341. M.G. Zelin, A.K. Mukherjee: Analysis of the cooperative grain boundary sliding in terms of cellular dislocations, Phil. Mag. A **68**, No. 6, 1183–1193 (1993)
342. M.G. Zelin, M.R. Dunlap, R. Rosen, A.K. Mukherjee: The direct observation of cooperative grain-boundary sliding and migration during superplastic deformation of lead–tin eutectic in shear, J. Appl. Phys. **74**, No. 8, 4972–4982 (1993)
343. M.G. Zelin, A.K. Mukherjee: Microstructural aspects of non-homogeneity of grain-boundary sliding, J. Mater. Sci. **28**, No. 24, 6767–6773 (1993)
344. M.G. Zelin, N.A. Krasilnikov, R.Z. Valiev, M.W. Grabski, H.S. Yang, A.K. Mukherjee: Microstructural aspects of the nonhomogeneity of superplastic deformation at the level of grain groups, Acta Metall. Mater. **42**, No. 1, 119–126 (1994)
345. M.G. Zelin, A.K. Mukherjee: Cooperative phenomena at grain boundaries during superplastic flow, Acta Metall. Mater. **43**, No. 6, 2359–2372 (1995)
346. M.G. Zelin, A.K. Mukherjee: High temperature deformation behavior of titanium samples with superplastic layer, J. Mater. Res. **10**, No. 4, 864–869 (1995)
347. J.E. Morral, M.F. Ashby: Dislocated cellular structures, Acta Metall. **22**, No. 5, 567–575 (1974)
348. M.G. Zelin, S. Guillard, A.K. Mukherjee: Cellular dislocations: Experiment and modeling, Mater. Sci. Eng. A **309–310**, 514–519 (2001)
349. J.P. Hirth, J. Lothe: *Theory of Dislocations* (John Wiley, New York 1982)
350. W.D. Nix: *Rate Processes and Plastic Deformation in Materials*, Vol. 4, ed. by J.C.M. Li, A.K. Mukherjee (American Society of Metals, Metals Park, OH 1975) p. 384
351. M.Yu. Gutkin, I.A. Ovid'ko: On role of interfaces in plastic deformation processes in nanostructured boride and nitride films, J. Mater. Processing & Manufacturing Science **7**, No. 1, 59–65 (1998)
352. K.L. Johnson: *Contact Mechanics*, Chaps. 1–5 (Cambridge Univ. Press, Cambridge 1985)
353. V.Yu. Milman: Analysis of microhardness behavior, Acta Mater. **42**, 1349–1356 (1994)
354. R.D. Mindlin: Second gradient of strain and surface-tension in linear elasticity, Int. J. Solids Structures **1**, No. 4, 417–438 (1965)
355. S.B. Altan, E.C. Aifantis: On the structure of the mode III crack-tip in gradient elasticity, Scripta Metal. Mater. **26**, No. 2, 319–324 (1992)

356. E.C. Aifantis: Gradient effects at the micro-, macro- and nanoscales, J. Mech. Behaviour of Materials **5**, No. 3, 355–375 (1994)
357. B.S. Altan, E.C. Aifantis: On some aspects in the special theory of gradient elasticity, J. Mech. Behaviour of Materials **8**, No. 3, 231–282 (1997)
358. C.Q. Ru, E.C. Aifantis: A simple approach to solve boundary-value problems in gradient elasticity, Acta Mechanica **101**, No. 1, 59–68 (1993)
359. D.J. Unger, E.C. Aifantis: The asymptotic solution of gradient elasticity for mode III, Int. J. Fracture **71**, No. 2, R27–R32 (1995)
360. G.E. Exadaktylos, E.C. Aifantis: Two and three dimensional crack problems in gradient elasticity, J. Mech. Behaviour of Materials **7**, No. 1, 93–117 (1996)
361. I. Vardoulakis, G. Exadaktylos, E.C. Aifantis: Gradient elasticity with surface energy: Mode-III crack problem, Int. J. Solids Structures **33**, 4531–4559 (1996)
362. C.Q. Ru, E.C. Aifantis: *Some Studies on Boundary Value Problems in Gradient Elasticity* (Preprint of the MTU Report, Houghton, MI 1993) unpublished
363. W.W. Milligan, S.A. Hackney, E.C. Aifantis: Constitutive modeling for nanostructured materials. In: *Continuum Models for Materials with Microstructure*, ed. by H. Mühlhaus (Wiley 1995) pp. 379–393
364. R. de Wit: Theory of disclinations: IV. Straight disclinations, J. Res. Nat. Bur. Stand. (U.S.) **77**A, 608–658 (1973)
365. M.Yu. Gutkin, E.C. Aifantis: Unpublished results (2003)
366. G. Gutekunst, J. Mayer, M. Rühle: Atomic structure of the interface between epitaxial niobium films and α-Al_2O_3 substrates, Mater. Sci. Forum **207**–**209**, Part 1, 241–244 (1996)
367. P.C. Gehlen, A.R. Rosenfield, G.T. Hahn: Structure of the ⟨100⟩ edge dislocation in iron, J. Appl. Phys. **39**, No. 11, 5246–5254 (1968)
368. M.Yu. Gutkin, I.A. Ovid'ko: Kinetics of cellular dislocations in nanostructured solids under superplastic deformation, Ann. Chim. Sci. Mat. **27**, No. 3, 111–116 (2002)
369. V.F. Zaitsev, A.D. Polyanin: *Handbook of Ordinary Differential Equations: Exact Solutions* (Fizmatlit, Moscow 1995) in Russian
370. R.S. Mishra, R.Z. Valiev, S.X. McFadden, A.K. Mukherjee: Tensile superplasticity in a nanocrystalline nickel aluminide, Mater. Sci. Eng. A **252**, No. 2, 174–178 (1998)
371. R.K. Islamgaliev, R.Z. Valiev, R.S. Mishra, A.K. Mukherjee: Enhanced superplastic properties in bulk metastable nanostructured alloys, Mater. Sci. Eng. A **304**–**306**, 206–210 (2001)
372. R.S. Mishra, V.V. Stolyarov, C. Echer, R.Z. Valiev, A.K. Mukherjee: Mechanical behavior and superplasticity of a severe plastic deformation processed nanocrystalline Ti–6Al–4V alloy, Mater. Sci. Eng. A **298**, Nos. 1–2, 44–50 (2001)
373. R.Z. Valiev, C. Song, S.X. McFadden, A.K. Mukherjee, R.S. Mishra: TEM/HREM observations of nanostructured superplastic Ni_3Al, Phil. Mag. A **81**, No. 1, 25–36 (2001)
374. R.Z. Valiev, I.V. Alexandrov, Y.T. Zhu, T.C. Lowe: Paradox of strength and ductility in metals processed by severe plastic deformation, J. Mater. Res. **17**, No. 1, 5–8 (2002)
375. A.K. Mukherjee: Elevated temperature crystalline plasticity at diminishing length scales. In: *Creep Deformation: Fundamentals and Applications*, ed. by R.S. Mishra, J.C. Earthman, S.V. Raj (TMS, Warrendale 2002) pp. 3–19

376. M.Yu. Gutkin, I.A. Ovid'ko, N.V. Skiba: Strengthening mechanism for high-strain-rate superplasticity in nanocrystalline materials, J. Phys. D: Appl. Phys. **36**, No. 12, L47–L50 (2004)
377. M.Yu. Gutkin, I.A. Ovid'ko, N.V. Skiba: Strengthening and softening mechanisms in nanocrystalline materials under superplastic deformation, Acta Mater. (2003)
378. T. Mura: The continuum theory of dislocations. In: *Advances in Materials Research*, Vol. 3, ed. by H. Herman (Interscience Publishers, New York London Sydney Toronto 1968) pp. 1–108
379. H. Hahn, K.A. Padmanabhan: A model for the deformation of nanocrystalline materials, Philos. Mag. B **76**, No. 4, 559–571 (1997)
380. Q. Wei, D. Jia, K.T. Ramesh, E. Ma: Evolution and microstructure of shear bands in nanostructured Fe, Appl. Phys. Lett. **81**, No. 7, 1240–1242 (2002)
381. P. Klimanek, V. Klemm, A.E. Romanov, M. Seefeldt: Disclinations in plastically deformed metallic materials, Adv. Eng. Mater. **3**, No. 11, 877–884 (2001)
382. U. Essmann: Die Versetzunngsanordnung in plastisch verformten Kupfereinkristallen, Phys. Stat. Sol. **3**, No. 5, 932–949 (1963)
383. J.W. Steeds: Dislocation arrangement in copper single crystals as a function of strain, Proc. Roy. Soc. A **292**, No. 1430, 343–373 (1966)
384. A.N. Vergazov, V.A. Likhachev, V.V. Rybin: Characteristic elements of dislocation structure in deformed polycrystalline molybdenum, Fizika Metallov i Metallovedenie **42**, No. 1, 146–154 (1976) in Russian [English translation in: Phys. Metals Metallogr.]
385. G.V. Berezhkova, P.P. Perstnev: Effect of high-temperature prestrain on plastic properties of magnesium oxide crystals, Doklady AN SSSR **248**, No. 5, 1105–1108 (1979) in Russian [English translation in: Sov. Phys. Dokl. **24**, No. 10, 799–800 (1979)]
386. G.V. Berezhkova, P.P. Perstnev, A.E. Romanov, V.I. Vladimirov: Peculiarities of reoriented bands formation in crystals, Cryst. Res. Techn. **18**, No. 2, 139–147 (1983)
387. Ye.V. Nesterova, V.V. Rybin: Mechanical twinning and fragmentation of technically pure titanium on developed plastical deformation stage, Fizika metallov i metallovedenie **59**, No. 2, 395–406 (1985) in Russian [English translation in: Phys. Met. Metall.]
388. A.D. Korotaev, A.N. Tyumentsev, V.F. Sukhovarov: *Dispersion Hardening of Refractory Metals* (Nauka, Novosibirsk 1989) in Russian
389. A.D. Korotaev, A.N. Tyumentsev, Yu.P. Pinjin: Activation and characteristic types of defect substructures at mesolevel of plastic flow in high-strength materials, Fizicheskaya mezomekhanika **1**, No. 1, 23–35 (1998) in Russian
390. M.Yu. Gutkin, K.N. Mikaelyan, A.E. Romanov, P. Klimanek: Disclination models of misorientation band generation and propagation, Phys. Stat. Sol. (a) **193**, No. 1, 35–52 (2002)
391. V.I. Vladimirov, A.E. Romanov: Partial disclination dipole motion under plastic deformation, Fizika tverdogo tela **20**, No. 10, 3114–3116 (1978) in Russian [English translation in: Sov. Phys. Solid State **20**, No. 10, 1795–1796 (1978)]
392. B.K. Barakhtin, V.I. Vladimirov, S.A. Ivanov, I.A. Ovid'ko, A.E. Romanov: Periodic structural changes during rotational plastic deformation, Fizika metallov i metallovedenie **63**, No. 6, 1185–1191 (1987) in Russian [English translation in: Phys. Met. Metall. **63**, No. 6, 133–139 (1987)]

393. A.E. Romanov, E.C. Aifantis: On the kinetic and diffusional nature of linear defects, Scr. Metall. Mater. **29**, No. 4, 707–712 (1993)
394. M. Seefeldt, P. Klimanek: Interpretation of plastic deformation by means of dislocation–disclination reaction kinetics, Mater. Sci. Eng. A **234–236**, 758-761 (1997)
395. M. Seefeldt, P. Klimanek: Modelling of flow behaviour of metals by means of a dislocation–disclination reaction kinetics, Model. Simul. Mater. Sci. Eng. **6**, No. 4, 349–360 (1998)
396. K.N. Mikaelyan, M. Seefeldt, M.Yu. Gutkin, P. Klimanek, A.E. Romanov: Computer simulation of 2D dislocation–disclination dynamics, Fizika tverdogo tela **45**, No. 11, 2002–2007 (2003) in Russian [English translation in: Phys. Solid State **45**, No. 11, 2104–2109 (2003)]
397. V.A. Likhachev, R.Yu. Khairov: *Introduction to the Theory of Disclinations* (Leningrad State University, Leningrad 1975) in Russian
398. R. de Wit: Theory of disclinations: IV. Straight disclinations, J. Res. Nat. Bur. Stand. **77**A, No. 5, 607–658 (1973)
399. A. Needleman: Computational mechanics at the nanoscale, Acta Mater. **48**, No. 1, 105–124 (2000)
400. B. Devincre, L.P. Kubin, C. Lemarchand, R. Madec: Mesoscopic simulations of plastic deformation, Mater. Sci. Eng. A **309–310**, 211–219 (2001)
401. L. Nicola, E. Van der Giessen, A. Needleman: 2D dislocation dynamics in thin metal layers, Mater. Sci. Eng. A **309–310**, 274–277 (2001)
402. N. Argaman, O. Levy, G. Makov: When do 2-D dislocations form cellular structures? Mater. Sci. Eng. A **309–310**, 386–392 (2001)
403. O. Politano, J.M. Salazar: A 3D mesoscopic approach for discrete dislocation dynamics, Mater. Sci. Eng. A **309–310**, 261–264 (2001)
404. H. Yasin, H.M. Zbib, M.A. Khaleel: Size and boundary effects in discrete dislocation dynamics: Coupling with continuum finite element, Mater. Sci. Eng. A **309–310**, 294–299 (2001)
405. A.A. Nazarov, A.E. Romanov, R.Z. Valiev: Random disclination ensembles in ultrafine-grained materials produced by severe plastic deformation, Scr. Mater. **34**, No. 5, 729–734 (1996)
406. M.F. Ashby: The deformation of plastically non-homogeneous material, Phil. Mag. **21**, No. 170, 399–424 (1970)
407. U.F. Kocks, A.S. Argon, M.F. Ashby: Thermodynamics and kinetics of slip, Progr. Mater. Sci. **19**, 1–288 (1975)
408. K.M. Jassby, T. Vreeland Jr.: An experimental study of the mobility of edge dislocations in pure copper single crystals, Phil. Mag. **21**, No. 174, 1147–1168 (1970)
409. R.Z. Valiev, T.G. Langdon: An investigation of the role of intragranular dislocation strain in the superplastic lead-62% tin eutectic alloy, Acta Metall. **41**, No. 3, 949–954 (1993)
410. M. Seefeldt, P. Van Houtte: A disclination-based model for anisotropic substructure development and its impact on the critical resolved shear stresses, Mater. Phys. Mech. **1**, No. 2, 133–139 (2000)
411. M.Yu. Gutkin, I.A. Ovid'ko: Transformations of low-angle tilt boundaries in high-T_c Superconductors, Phys. Rev. B **63**, No. 6, 064515 (2001)
412. S.V. Bobylev, I.A. Ovid'ko, A.G. Sheinerman: Effects of misfit stresses on the structure and transport properties of grain boundaries in high-T_c superconducting films, Phys. Rev. B **64**, No. 22, 224507 (2001)

413. N.I. Noskova, E.G. Volkova: In situ study of deformation of nanocrystalline copper, Fizika Metallov i Metallovedenie **91**, No. 6, 100–107 (2001) in Russian [English translation in: Phys. Met. Metall. **91**, No. 6, 629–635 (2001)]
414. N.I. Noskova, E.G. Volkova: In situ investigations of deformation in nanocrystalline, $Fe_{73.5}Cu_1Nb_3Si_{13.5}B_9$ alloy, Fizika Metallov i Metallovedenie **92**, No. 4, 107–111 (2001) in Russian [English translation in: Phys. Met. Metall. **92**, No. 4, 421–425 (2001)]
415. N.I. Noskova: Physics of deformation of nanocrystalline metals and alloys. In: *Problems of Nanocrystalline Materials*, ed. by V.V. Ustinov, N.I. Noskova (Ural Department of Russian Academy of Sciences, Ekaterinburg 2002) pp. 159–170, in Russian
416. K.E. Harris, V.V. Singh, A.H. King: Grain rotation in thin films of gold, Acta Mater. **46**, No. 8, 2623–2633 (1998)
417. P. Müllner, W.-M. Kuschke: Disclinations due to grain boundary relaxation in fine-grained materials and thin films, Scr. Mater. **36**, No. 12, 1451–1455 (1997)
418. I.A. Ovid'ko: Deformation of nanostructures, Science **295**, 2386 (2002)
419. M.Yu. Gutkin, A.L. Kolesnikova, I.A. Ovid'ko, N.V. Skiba: Rotational deformation in fine-grained materials prepared by severe plastic deformation, J. Metastable & Nanostruct. Mater. **12**, 47–58 (2002)
420. M.Yu. Gutkin, A.L. Kolesnikova, I.A. Ovid'ko, N.V. Skiba: Disclinations and rotational deformation in fine-grained materials, Phil. Mag. Lett. **82**, No. 12, 651–657 (2002)
421. M.Yu. Gutkin, I.A. Ovid'ko, N.V. Skiba: Changes in grain boundary misorientation caused by emission of dislocation pairs, Pis'ma v ZhTF **28**, No. 10, 78–82 (2002) in Russian [English translation in: Tech. Phys. Lett. **28**, No. 5, 437–438 (2002)]
422. M.Yu. Gutkin, I.A. Ovid'ko, N.V. Skiba: Transformations of grain boundaries due to disclination motion and emission of dislocation pairs, Mater. Sci. Eng. A **339**, 73–80 (2003)
423. H. Grimmer, W. Bollmann, D.H. Warrington: Coincidence-site lattices and complete pattern-shift lattices in cubic crystals, Acta Crystall. A **30**, 197–207 (1974)
424. A.P. Sutton, R.W. Balluffi: *Interfaces in Crystalline Materials* (Clarendon Press, Oxford 1995)
425. M.Yu. Gutkin, K.N. Mikaelyan, I.A. Ovid'ko: Linear splitting of disclinations in polycrystalline and nanocrystalline solids, Fizika tverdogo tela **37**, No. 2, 552–554 (1995) in Russian [English translation in: Sov. Phys. Solid State **37**, No. 2, 300–301 (1995)]
426. M.Yu. Gutkin, I.A. Ovid'ko, K.N. Mikaelyan: On the role of disclinations in relaxation and deformation processes in nanostructured materials, Nanostruct. Mater. **6**, Nos. 5–8, 779–782 (1995)
427. M.Yu. Gutkin, K.N. Mikaelyan, I.A. Ovid'ko: Low-energy disclination structures at grain boundaries in polycrystalline and nanocrystalline solids, Phys. Stat. Sol. (a) **153**, No. 2, 337–346 (1996)
428. M.Yu. Gutkin, I.A. Ovid'ko, N.V. Skiba: Crossover from grain boundary sliding to rotational deformation in nanocrystalline materials, Acta Mater. **51**, No. 14, 4059–4071 (2003)
429. J.D. Eshelby, F.C. Frank, F.R.N. Nabarro: The equilibrium of linear arrays of dislocations, Phil. Mag. **42**, No. 327, 351–364 (1951)

430. V.I. Vladimirov: *Physical Nature of Fracture of Metals* (Metallurgiya, Moscow 1984) in Russian
431. V.V. Nemoshkalenko (Ed.): *Cooperative Deformation Processes and Localization of Deformation* (Naukova Dumka, Kiev 1989) in Russian
432. V.E. Panin, V.A. Likhachev, Yu.V. Grinyaev: *Structural Levels of Deformation in Solids* (Nauka, Novosibirsk 1985) in Russian
433. V.E. Panin (Ed.): *Structural Levels of Plastic Deformation and Fracture* (Nauka, Novosibirsk 1990) in Russian
434. V.E. Panin (Ed.): *Physical Meso-Mechanics and Computer Design of Materials*, Vols. 1–2 (Nauka, Novosibirsk 1995) in Russian
435. V.V. Rybin, A.A. Zisman: Structural micromechanics of plastic deformation and viscous fracture of fragmented crystals, Problemy prochnosti **3**, 70–77 (1985) in Russian
436. N.A. Koneva, D.V. Lychagin, L.A. Teplyakova, E.V. Kozlov: Rotations of crystalline lattice and stages of plastic deformation. In: *Experimental Investigation and Theoretical Description of Disclinations*, ed. by V.I. Vladimirov (Ioffe Physico-Technical Institute, Leningrad 1984) pp. 161–167, in Russian
437. N.A. Koneva, D.V. Lychagin, L.A. Teplyakova, E.V. Kozlov: Dislocation disclination substructures and strengthening. In: *Theoretical and Experimental Investigation of Disclinations*, ed. by V.I. Vladimirov (Ioffe Physico-Technical Institute, Leningrad 1986) pp. 116–126, in Russian
438. N.A. Koneva, D.V. Lychagin, L.I. Trishkina, E.V. Kozlov: Accumulation of defects, stored strain energy and self-organization of substructure. In: *Physical Aspects of Prognostication for Fracture and Deformation of Heterogeneous Materials*, ed. by A.M. Leksovskii (Ioffe Physico-Technical Institute, Leningrad 1987) pp. 20–36, in Russian
439. N.A. Koneva, D.V. Lychagin, L.A. Teplyakova, L.I. Trishkina, E.V. Kozlov: Streaky substructure in FCC single phase alloys. In: *Disclinations and Rotational Deformation of Solids*, ed. by V.I. Vladimirov (Ioffe Physico-Technical Institute, Leningrad 1988) pp. 103 113, in Russian
440. E.V. Kozlov, N.A. Koneva, L.I. Trishkina: Dislocation–disclination substructures and bending–twisting of crystalline lattice. In: *Disclinations and Rotational Deformation of Solids*, ed. by A.E. Romanov (Ioffe Physico-Technical Institute, Leningrad 1990) pp. 89–125, in Russian
441. V.A. Pavlov: Amorphization of structure in metals and alloys with limitingly high grade of deformation, Fizika metallov i metallovedenie **59**, No. 4, 629–649 (1986) in Russian [English translation in: Phys. Met. Metall.]
442. V.A. Pavlov, M.V. Popov: Relation between the processes of fracture and amorphization of structure of crystalline systems under cold plastic deformation, Fizika metallov i metallovedenie **2**, 192–198 (1990) in Russian [English translation in: Phys. Met. Metall.]
443. V.M. Segal, V.I. Reznikov, V.I. Kopylov, D.A. Pavlik, V.F. Malyshev: *Processes of Plastic Formation of Metal Structure* (Navuka i Tekhnika, Minsk 1994) in Russian
444. M.Yu. Gutkin, I.A. Ovid'ko: Nanocracks at grain boundaries in nanocrystalline materials, submitted to Phil. Mag. Lett. (2003)
445. I.V. Zolotukhin: *Physical Properties of Amorphous Metallic Materials* (Metallurgiya, Moscow 1986) in Russian

446. A.M. Glezer, B.V. Molotilov: Structure of amorphous alloys, Fizika metallov i metallovedenie **2**, 5–28 (1990) in Russian [English translation in: Phys. Met. Metall.]
447. V.A. Likhachev, V.V. Rybin: Disclinations in an ideally fragmented crystal, Fizika tverdogo tela **18**, No. 1, 163–165 (1976) in Russian [English translation in: Sov. Phys. Solid State **18**, No. 1, 93–94 (1976)]
448. V.V. Rybin, I.M. Zhukovskii: Disclination mechanism of microcrack formation, Fizika tverdogo tela **20**, No. 6, 1829–1835 (1978) in Russian [English translation in: Sov. Phys. Solid State **20**, No. 6, 1056–1059 (1978)]
449. W. Bollmann: Triple lines in polycrystalline aggregates as disclinations, Phil. Mag. A **49**, No. 1, 73–79 (1984)
450. W. Bollmann: Triple-line disclinations representations, continuity and reactions, Phil. Mag. A **57**, No. 4, 637–649 (1988)
451. W. Bollmann: Triple-line disclinations in polycrystalline material, Mater. Sci. Eng. A **113**, No. 1, 129–138 (1989)
452. V.V. Rybin, A.A. Zisman, N.Yu. Zolotorevsky: Junction disclinations in plastically deformed crystals, Acta Metall. Mater. **41**, 2211–2217 (1993)
453. S.G. Zaichenko, A.V. Shalimova, A.O. Titov: Investigation of triple junction defect structure, Mater. Sci. Forum **207–209**, Part 1, 373–376 (1996)
454. I.A. Ovid'ko, A.V. Osipov: Splitting of disclinations and the 'crystal–glass' transition in mechanical alloying, Fizika tverdogo tela **34**, No. 1, 288–292 (1992) in Russian [English translation in: Phys. Solid State **34**, No. 1, 152–154 (1992)]
455. V.A. Pavlov: High plastic deformations and nature of amorphization and dispercing of crystalline systems, Fizika metallov i metallovedenie **67**, No. 5, 924–944 (1989) in Russian [English translation in: Phys. Met. Metall.]
456. S.G. Zaichenko, A.V. Shalimova, A.O. Titov, A.M. Glezer: Role of triple junctions in developed plastic deformation of polycrystals. In: *Relaxational Phenomena in Defect Structures of Solids: Proceedings of International Seminar (Voronej, 5–8 September 1995)*, Part II (Voronej State Technical University, Voronej 1996) pp. 212–224, in Russian
457. V.L. Indenbom: On criteria of fracture in dislocation theories of strength, Fizika tverdogo tela **3**, No. 7, 2071–2079 (1961) in Russian [English translation in: Sov. Phys. Solid State **3**, No. 7, 1506–1511 (1961)]
458. V.V. Rybin, A.N. Vergazov, V.A. Likhachev: Viscous fracture due to fragmentation of structure, Fizika metallov i metallovedenie **37**, No. 3, 620–624 (1974) in Russian [English translation in: Phys. Met. Metall.]
459. V.R. Regel, A.M. Leksovskii, S.N. Sakiev: The kinetics of the thermofluctuation-induced micro- and macrocrack growth in plastic metals, Int. J. Fracture **11**, No. 5, 841–850 (1975)
460. R.I. Lyles, H.G.F. Wilsdorf: Microcrack nucleation and fracture in silver crystals, Acta Metall. **23**, No. 2, 269–277 (1975)
461. E. Pippel, J. Woltersdorf: High-voltage and high-resolution electron microscopy studies of interfaces in zirconia-toughened alumina, Phil. Mag. A **56**, No. 5, 595–613 (1987)
462. S. Veprek, A.S. Argon: Towards the understanding of mechanical properties of super- and ultrahard nanocomposites, J. Vac. Sci. Technol. **20**, No. 2, 650–664 (2002)
463. A.P. Prudnikov, Yu.A. Brychkov, O.I. Marichev: *Integrals and Series. Elementary Functions* (Nauka, Moscow 1981) in Russian
464. J.R. Weertman, P.G. Sanders: Plastic deformation of nanocrystalline metals, Solid State Phenom. **35–36**, 249–262 (1994)

Springer Series in
MATERIALS SCIENCE

Editors: R. Hull R. M. Osgood, Jr. J. Parisi H. Warlimont

10 **Computer Simulation of Ion-Solid Interactions**
By W. Eckstein

11 **Mechanisms of High Temperature Superconductivity**
Editors: H. Kamimura and A. Oshiyama

12 **Dislocation Dynamics and Plasticity**
By T. Suzuki, S. Takeuchi, and H. Yoshinaga

13 **Semiconductor Silicon**
Materials Science and Technology
Editors: G. Harbeke and M. J. Schulz

14 **Graphite Intercalation Compounds I**
Structure and Dynamics
Editors: H. Zabel and S. A. Solin

15 **Crystal Chemistry of High-T_c Superconducting Copper Oxides**
By B. Raveau, C. Michel, M. Hervieu, and D. Groult

16 **Hydrogen in Semiconductors**
By S. J. Pearton, M. Stavola, and J. W. Corbett

17 **Ordering at Surfaces and Interfaces**
Editors: A. Yoshimori, T. Shinjo, and H. Watanabe

18 **Graphite Intercalation Compounds II**
Editors: S. A. Solin and H. Zabel

19 **Laser-Assisted Microtechnology**
By S. M. Metev and V. P. Veiko
2nd Edition

20 **Microcluster Physics**
By S. Sugano and H. Koizumi
2nd Edition

21 **The Metal-Hydrogen System**
By Y. Fukai

22 **Ion Implantation in Diamond, Graphite and Related Materials**
By M. S. Dresselhaus and R. Kalish

23 **The Real Structure of High-T_c Superconductors**
Editor: V. Sh. Shekhtman

24 **Metal Impurities in Silicon-Device Fabrication**
By K. Graff 2nd Edition

25 **Optical Properties of Metal Clusters**
By U. Kreibig and M. Vollmer

26 **Gas Source Molecular Beam Epitaxy**
Growth and Properties of Phosphorus Containing III–V Heterostructures
By M. B. Panish and H. Temkin

27 **Physics of New Materials**
Editor: F. E. Fujita 2nd Edition

28 **Laser Ablation**
Principles and Applications
Editor: J. C. Miller

29 **Elements of Rapid Solidification**
Fundamentals and Applications
Editor: M. A. Otooni

30 **Process Technology for Semiconductor Lasers**
Crystal Growth and Microprocesses
By K. Iga and S. Kinoshita

31 **Nanostructures and Quantum Effects**
By H. Sakaki and H. Noge

32 **Nitride Semiconductors and Devices**
By H. Morkoç

33 **Supercarbon**
Synthesis, Properties and Applications
Editors: S. Yoshimura and R. P. H. Chang

34 **Computational Materials Design**
Editor: T. Saito

35 **Macromolecular Science and Engineering**
New Aspects
Editor: Y. Tanabe

36 **Ceramics**
Mechanical Properties, Failure Behaviour, Materials Selection
By D. Munz and T. Fett

37 **Technology and Applications of Amorphous Silicon**
Editor: R. A. Street

38 **Fullerene Polymers and Fullerene Polymer Composites**
Editors: P. C. Eklund and A. M. Rao

Springer Series in
MATERIALS SCIENCE

Editors: R. Hull R. M. Osgood, Jr. J. Parisi H. Warlimont

39 **Semiconducting Silicides**
Editor: V. E. Borisenko

40 **Reference Materials in Analytical Chemistry**
A Guide for Selection and Use
Editor: A. Zschunke

41 **Organic Electronic Materials**
Conjugated Polymers and Low Molecular Weight Organic Solids
Editors: R. Farchioni and G. Grosso

42 **Raman Scattering in Materials Science**
Editors: W. H. Weber and R. Merlin

43 **The Atomistic Nature of Crystal Growth**
By B. Mutaftschiev

44 **Thermodynamic Basis of Crystal Growth**
P–T–X Phase Equilibrium and Non-Stoichiometry
By J. Greenberg

45 **Thermoelectrics**
Basic Principles and New Materials Developments
By G. S. Nolas, J. Sharp, and H. J. Goldsmid

46 **Fundamental Aspects of Silicon Oxidation**
Editor: Y. J. Chabal

47 **Disorder and Order in Strongly Nonstoichiometric Compounds**
Transition Metal Carbides, Nitrides and Oxides
By A. I. Gusev, A. A. Rempel, and A. J. Magerl

48 **The Glass Transition**
Relaxation Dynamics in Liquids and Disordered Materials
By E. Donth

49 **Alkali Halides**
A Handbook of Physical Properties
By D. B. Sirdeshmukh, L. Sirdeshmukh, and K. G. Subhadra

50 **High-Resolution Imaging and Spectrometry of Materials**
Editors: F. Ernst and M. Rühle

51 **Point Defects in Semiconductors and Insulators**
Determination of Atomic and Electronic Structure from Paramagnetic Hyperfine Interactions
By J.-M. Spaeth and H. Overhof

52 **Polymer Films with Embedded Metal Nanoparticles**
By A. Heilmann

53 **Nanocrystalline Ceramics**
Synthesis and Structure
By M. Winterer

54 **Electronic Structure and Magnetism of Complex Materials**
Editors: D.J. Singh and D. A. Papaconstantopoulos

55 **Quasicrystals**
An Introduction to Structure, Physical Properties and Applications
Editors: J.-B. Suck, M. Schreiber, and P. Häussler

56 **SiO_2 in Si Microdevices**
By M. Itsumi

57 **Radiation Effects in Advanced Semiconductor Materials and Devices**
By C. Claeys and E. Simoen

58 **Functional Thin Films and Functional Materials**
New Concepts and Technologies
Editor: D. Shi

59 **Dielectric Properties of Porous Media**
By S.O. Gladkov

60 **Organic Photovoltaics**
Concepts and Realization
Editors: C. Brabec, V. Dyakonov, J. Parisi and N. Sariciftci

MIX
Papier aus verantwortungsvollen Quellen
Paper from responsible sources
FSC® C105338

If you have any concerns about our products,
you can contact us on
ProductSafety@springernature.com

In case Publisher is established outside the EU,
the EU authorized representative is:
Springer Nature Customer Service Center GmbH
Europaplatz 3, 69115 Heidelberg, Germany

Printed by Libri Plureos GmbH
in Hamburg, Germany